I0007933

Arduino

Getting Started With Arduino and Basic
Programming With Projects

*(Advanced Methods to Learn Arduino
Programming)*

Ernest Leclerc

Published By **Zoe lawson**

Ernest Leclerc

All Rights Reserved

*Arduino: Getting Started With Arduino and Basic
Programming With Projects (Advanced Methods
to Learn Arduino Programming)*

ISBN   978-1-77485-489-1

Legal & Disclaimer

The information contained in this ebook is not designed to replace or take the place of any form of medicine or professional medical advice. The information in this ebook has been provided for educational & entertainment purposes only.

The information contained in this book has been compiled from sources deemed reliable, and it is accurate to the best of the Author's knowledge; however, the Author cannot guarantee its accuracy and validity and cannot be held liable for any errors or omissions. Changes are periodically made to this book. You must consult your doctor or get professional medical advice before using any of the suggested remedies, techniques, or information in this book.

TABLE OF CONTENTS

Introduction

The chapters to follow will go over all the various components that we must know before we can use Arduino. Arduino microcontroller.

There are plenty of devices that we can to use when it comes time to tackle our work in the field of technology and learn to programme, but none is capable of providing the same benefits and advantages that we get from working using Arduino system.

This guidebook will look into this controller and take an review of some of the ways in which we can use it to meet our requirements.

This guidebook will provide a glimpse of the fundamentals we can handle with regard to using the Arduino controller.

We'll take a look deeper into some of the history that have been made using the Arduino controller, the features of it is, how is using it and some of the benefits of using this controller in general.

Then , we can shift our attention to the most important concepts we should pay attention to, for instance, the structure of our board to ensure that we make use of this in the correct way.

After all that's completed, it's time to look through and figure out how we can set up our Arduino board to use it to meet our requirements.

We'll be studying the best way to choose a board, the best way to run Arduino and how to perform some of the programming tasks and much more.

This leads us straight into the fundamentals of programming in Arduino like the syntax, structure, and the various types of data before we get into the fundamentals of making this controller the device we'd like to utilize.

We'll also talk some details about C as a C languages in the manual and how we can be in a position to use it to aid us in our progress.

There are plenty of possibilities to explore when it comes to programming on this board, but the C language is likely to be one of the most effective alternatives due to its user-friendliness and the fact that it's very similar to the Arduino language we'll examine in addition.

Then, we're going look at some of the other things we can achieve using this language and the controller.

We will discuss how to create logic statements, how we can manage our own operators as well as how to handle some of the tasks we have to manage when working with computers with the various API functions, as well as how to interact with the streaming class of this controller.

Then, we can conclude this guidebook with a discussion on how to design the user-defined functions we want to use to ensure that we benefit from our brand new software for coding.

There are many various components that we can work on when it comes to Arduino's controller, and all the amazing things it can do to do.

If we're ready to know what the Arduino controller and Arduino board perform and the various things we can do with it, we should make sure we go through this manual to help begin.

There are many books on the subject available out there, so thanks to you for choosing this one!

All efforts were taken to ensure that it's filled with as much helpful information as it is possible. have fun!

Chapter 1: The Basics Of Understanding Arduino

We live living in an era of technology at the moment and there's always an increase in growing levels of technological literacy that even the common person should possess significantly more than was expected in the past.

A growing number of people are learning about how to make use of the array of hardware and software that are in the market. There is an increase in interest for them to tackle one or two of these items regardless of whether they intend to work on it in the professional realm or as an interest.

In these times there is a tendency to see a large number of novice programmers to become drawn by the possibilities we have on Arduino. Arduino system.

Perhaps you've seen a handful of initiatives that this system could perform on behalf of us. have been lucky enough to learn more about it and the ways that its versatility and simplicity could make life a more manageable in general.

No matter what the situation is that you are facing, you will discover that there are lots of advantages that can be derived from making use of the Arduino system for your personal requirements and needs.

With these concepts in mind, it is important to go deeper into the details of exactly what Arduino system is about and how we can work with the various technologies and how we're capable of making all of these provide the benefits we're looking for when we use it.

Before that, however we should take a look at the history behind this technology.

The Arduino History

The first step is to take a closer look at the stories that are set to be created by this Arduino device.

The technology that was developed in this was initially an idea in 2003. Hernando Barragan was looking to create an idea that could streamline the process by which could work. BASIC stamp microcontroller to operate, while making sure there was a decrease in the expenses associated with it.

This will make it much easier for engineers and students to acquire and use this technology without the costs.

For the first one the Arduino microcontroller is will be a smaller component connected to a computer which can be programmed and work with to accomplish various functions we require in the present.

In 2003, the majority of these controllers are priced around $100. usually, they are higher if you want to add some extra features that can handle it.

However, this isn't the case when working using Arduino. Arduino system.

These will reduce many of the expenses you'll observe with these controllers, without costing you a fortune.

The technology was created and refined over time, and in 2013 there was around 700,000 microcontroller boards available from Adafruit

Industries alone, a New York City supplier of these boards.

There were some problems with the trademarking process of Arduino that ended up with a division within Arduino for couple of months, Arduino is now going to be a singular company that is devoted to the creation of hardware and software that is accessible to people of all ages, but being flexible enough in the design to draw in engineers of a higher level.

What is Arduino?

This will lead us back to what exactly the Arduino board is about.

The background of Arduino is a little difficult at present However, the positive aspect to remember is that it's not going to be nearly as complex as it appears at the moment.

It is possible to look through this guide and find out more about the system as a whole place, what it's capable of doing to us and much more when we read through the entire guidebook.

The first aspect which we'll examine involves how this Arduino board will function.

First, you'll need to buy the equipment, selecting the option that best aligns with your objectives for this project.

Then, with the assistance of a machine that runs another major operating system no matter if we're discussing Mac, Linux, or Windows you will have to be spending time writing your codes in the correct format, with instructions for the board, uploading the entire sequence of instructions and steps you would like for the board to follow the codes you're writing in the process.

The code you create during this procedure will be saved within the microcontroller and will operate based on these instructions at any time you invoke them.

The great thing about this type of boards is you're capable of writing all the directions you'd like to and they'll go to follow them.

You can make a sound at the moment you turn on the controller, or have it beep when it detects that there is a light.

It's a great method to learn the fundamentals of coding throughout the process There are many options you're truly able to investigate to achieve this.

Who Can Use Arduino?

You might be amazed by the many people who use the Arduino board, and benefit from it to meet their requirements.

There are plenty of people who use the controller to tackle their projects, regardless of whether they can be for work, or just as an interest.

Some professionals who are interested in working with this board as well. Since all the programming in this article is easy and easy to use for the beginner and is rich enough that the novice is able to master a few skills and progress to more complex things later on. this particular board is gaining popularity in recent years.

It won't take time to get this system working before you realize that there are lots of people who want working with it.

Students and teachers alike love to use this product as well, and in the end it's going to be among the primary consumer bases for this type of product.

They were created to be a cost-effective technique that will assist us in building some of our own science instruments, so that teachers and students can perform exercises and demonstrations in physics, chemistry and many other areas also.

Furthermore, we'll be able to identify other people who are using this to meet their requirements too.

Architectural and designer can utilize this to create the models and prototypes that they'd like to build as time passes, allowing them to see what's happening as they modify and remove items.

Artists and musicians are capable of working with these to try out new instruments and techniques in accordance with what they want to accomplish in general.

The essence of what we're proposing here is that virtually everyone can use Arduino.

Even if you've got only the simplest of abilities in coding in any way, you will be able to see that this controller has been made for everyone and anyone to learn to utilize.

If you're looking to experiment with computer hardware or coding or simply want to understand the basics of microcomputers and software, the Arduino system will assist you in getting everything you need to know.

The Benefits of Arduino System Arduino System

The next thing we should be aware of is the benefits that we will see when we decide to utilize this method instead of one of the other.

There are many benefits we can observe when it comes time to implement this method.

This is the reason there are lots of programmers, whether novices, or even those who have been doing this for a long time and will use this microcontroller according to their personal requirements.

Some of the benefits and advantages we're capable of observing through this controller are the following:

1. The primary reason why these controllers were developed initially was the fact that they were efficient in terms of cost.

Instead of 100 or more on different boards, the tend to cost less than $50. Moreover, those boards that you build by yourself are likely to be at a cheaper price.

2. Arduino is the Arduino environment, also known as the IDE will run on a variety of different platforms in general.

This means you can make use of a computer with Windows on it just like the other boards that microcontrollers requires but it can also use an

Mac operating system system or any other computer running Linux can also be used.

This allows you to connect to any system you want to work using an IDE like the Arduino IDE.

3. The software you employ for Arduino will be classified as open-source.

The tools , or even the strings of code you'll utilize for instructing the controller what to do is something that anybody can access.

It is not necessary be through the procedure to purchase a licence order to utilize these tools, which makes it simpler for them to be utilized in the classroom.

4. The tools we're capable of using through the Arduino software could be expanded using C++ library as well as the AVR-C programming language.

This means that if you've got a more thorough knowledge of these languages you will be able to extend the technology and their capabilities more.

5. The environment you create to code for this control will be easy and straightforward to work with.

You'll be able to be aware of what the program is telling you to do and the actions that are required regardless of whether you're just beginning to learn about the subject.

This will make using this software significantly simpler in general.

6. The hardware is also open-sourced.

Anyone who has the motivation and knowledge could find and develop the hardware they would want to use and the Arduino program within the environment of the IDE.

Even those who aren't adept at creating circuits will be able to utilize the breadboard to design their own circuit boards using Arduino.

As we can see there are plenty of advantages we can work with making use of an Arduino system.

If you're looking for a microcontroller that'll allow everything done quickly and will allow you to get some great results from programming and also learning while doing it and not have to worry about the cost being too expensive to handle This is the perfect device for us to use.

What else should I know About Arduino?

Before we move to the final chapter, and then look at some of the other aspects we should to be aware of about Arduino it is important to be sure to cover some more aspects of this particular device so that we understand the reasons why it is extremely popular, and also why there are many young and old, trying to master how to utilize Arduino.

One thing we must keep in mind while working using Arduino is that it was designed to be low-cost in the back of our minds.

The reason was that other controllers on the market were extremely costly to operate and were somewhat of a struggle to master and test out.

Instead of Arduino board that costs priced at $100 or more to use, as the other boards,

Arduino is going to be a board that comes pre-assembled and priced at about $50.

If you're looking to capable of learning how to make this work by yourself You can purchase an entire board that includes all the components for less.

The IDE that is called the environment we can to work when working with Arduino will be a great choice because it will be used with other platforms that you're used to.

This means that it's not going to be a problem only with the Windows computer, even though that is the majority of microcontrollers available as well as computers with Linux and Mac also installed.

This allows the system appeal to more users since it's capable of allowing us to access the platform and operating system we'd like to use.

The third advantage or factor we should be working with is software is how it interacts in conjunction with Arduino. Arduino board.

This program will be considered open source so you can browse through it and utilize it at no cost.

All the code strings you need and that are utilized in this program will assist us tell the controller what you'd like it to accomplish and are also accessible for anyone wanting access to these strings.

This will be a benefit that you, as a programmers, will love since you won't have to waste lots of money on licensing to utilize the tools, as you would with other tools.

You may want to teach your students to utilize the Arduino board and these programs or to conduct the instruction on your own and get into coding it will be possible using Arduino. Arduino board, and without needing to think about all extra expenses that go along with it.

The open-sourced tools we've been discussing are going to be fairly easy to extend out if you choose to make use of the different C++ language and the other coding languages that fall in the category of AVR-C.

They'll be excellent libraries which include all the other features required to create code.

For those who are familiar with the basics of coding and the two main languages that are mentioned above, will discover that it is more simple to mix these two languages along with their features and libraries, to increase your knowledge on the subject.

It will also be a little more in the level of the program and its capabilities when you choose to make use of the C language. However, a person who is experienced and is prepared to commit more time to the process will find it easier to master, and is easy to master.

Do not fret if you're an absolute beginner.

These languages can assist you to realize what you want from of the code without any issues.

Another benefit of this is the environment that is created by Arduino.

This particular controller comes with an environment that is easy to use and clear to use.

This means that the application you are using and the IDE will be able handle certain instructions you transmit to your controller and will convert them into a format that is simple to use and doesn't need you to have years of experience or a higher level of education to be able to complete.

This is something other programmer had issues working with similar boards and this could hinder the kind of programming they'd like to achieve.

But, this isn't an issue with Arduino, and it won't take you long to are able to go on and actually work with the device, even as an amateur.

Finally, we have to have a clear idea about how hardware on this board will function.

The good thing is that this hardware is open-sourced.

The technology that comes together with this board are released under the Creative Commons License which is beneficial because it permits users, even when working on your coding to make the most basic modifications and changes which you'll need to make to make sure that the board functions without harm to be caused.

There's so much we can do using Arduino technology.

It might seem difficult at first but it's difficult to use or even comprehend If you've not been able to master programs in the past.

That's one of the many benefits that it offers.

It is an system and technology that was designed for every task you require but still make it easy for students to understand inside the schoolroom.

Even if you don't have a lot of knowledge of programming and coding or a technical background, you'll be able to learn this down.

The steps we cover in this guidebook can help in making this happen.

Chapter 2: Understanding About The Arduino

Board

After we have a few of the fundamentals regarding how to use the Arduino board , and the way it's designed to work, it's time to get an examination of what the board actually looks like and what it's supposed to appear.

If you glance at this display, it's likely to seem to make sense for a novice in programming. This could cause us to feel defeated before we even get the chance to begin.

That's why we're going to spend the next chapter studying at least a little concerning the Arduino board which is the actual board we will spend our time with, so that we know how to use it for our requirements before doing the programming.

An Overview of the Arduino Board

The first thing we should focus our attention on is it is the table it self.

There will be several components that come with the board. We need to know more of these components should we want to use the board in the future.

In this article, we'll begin by introducing some of the digital pins that you'll notice are located on the edges of the controller.

These will be vital because they're likely as inputs of our code or to sense out of conditions, and could aid in the output as the response that your controller is expected to make to the input it receives.

For instance an input we would like to hear could be similar to one of the lights of the sensor which go off when it's dark or when it detects that there is a deficiency of illumination.

This will result in it closing the circuit which is supposed to illuminate our bulb. Then this will be our output.

This is the scenario we'll be spending our time with when we are planning to work to transform the board into an evening light.

On the majority of boards we wish to work with, there's also an LED on the board, which will be linked to the pin we choose, like Pin 13 in the Uno of the Arduino.

This is the pin LED which will serve as the sole output option is inside the board and it's going to be present when we decide to make an illuminated light that flashes later.

The Pin LED will be an excellent tool to use whenever you'd like to investigate or correct some of the code that you've written. This will make sure that there aren't any types of errors that could appear in the process.

Its Power LED is going to be the exact thing we think of out of the word.

This will be the part that lights up when we can see that the board has been in operation or is receiving the power it is supposed to receive.

Another thing is that we can create using certain of the codes we are working on.

When you take a the time to look at the boards you'd prefer to utilize you will see that there will be the part that is called the microcontroller which is also known as the ATmega controller.

This will be the section which will manage the entire project and could be described as the brains of the project.

It will receive all the instructions your code will transmit to it and then act out according to the way it is in accordance with those codes and the instructions.

If you don't take the time to set up everything in place and then you don't, the board will encounter some problems when it comes time to begin working.

Then it's time to go to the section that will be all about analog pins.

Alongside the digital pins were just made it is necessary to use the analog pins that will be located close to the edges of our board in comparison with the pins digital that we created.

These will be the inputs that we can get from this system.

When we speak of the analog pins, it is discussing the signals which is expected to have an input, but it's not a continuous input, but one that has the ability to change each time.

This might be a thing that is different from an audio input employed by the system.

With this scenario in place the auditory signal in the room may vary a bit depending on the person who is within the space, the number of people are present in the room, and also other noises in the background.

The pins 5V and GND will be a different component that we must be aware of since they're utilized to supply extra power with 5V that is supplied for the circuit as well as our microcontroller.

The power connector you will find in this case will usually be located near the edge is visible on the Arduino board. It will be the one is used to provide energy to the microcontroller when it's not connected back to the USB.

It is possible to find that the USB port to be utilized to power our devices too, however generally speaking we will discover that the primary purpose of the USB port is in the transfer, or upload of the instructions we've coded on the computer in which you coded transfer onto the Arduino.

These TX as well as the RX lights will be used to to determine if there's an transfer of data which will take place in this instance.

This signal of the exchange is what we're going to see once we upload our sketch files from our computers to the Arduino to ensure that they will flash quickly as they perform the exchange.

We also have the reset button which will create a sound that will help reset the controller back to its default settings and wipe all the data we were capable of uploading to the Arduino.

Another Important Arduino Terms

Another thing we'll need to be aware of when working with an Arduino platform is various types of memory.

There are three kinds we can concentrate on with this kind of system.

Memory is any space in which we are able to keep the information needed to enable the system to function according to the way we desire.

The very first kind of memory we are in a position to work with is flash memory.

It is here that the source code for any program you write down is located.

It will be known as of program space since it will be utilized in an automated method for the program after we upload the program on the board.

This kind of memory will be in exactly the same location regardless of whether power is off or when it's time to turn off the board.

The second kind of memory we'll need to consider is referred to as static-random-access memory , or SRAM.

This will be the area located on our controller. your sketch or program you've created will be able to use to store, create and perform some work using the input sources in order to get the correct output from the process.

This type of storage is different, because If you don't save the data the data will disappear if you are unable to power the controller.

EEPROM will function similar to a small hard drive that allows the programmer to store data that is not related to the program once we've turned off the circuit.

There will be distinct instructions to assist in writing, reading, and erasing, as well as various other functions we'll require to perform.

There will be specific digital pins that will be tagged as PWM pins, in other words, they're being used to make an analog using certain digital methods.

Analog, as we need to be aware, implies that the input or output will be different and we're not going to be able to observe an exact pattern with this.

Most of the time, we're going to observe these digital pins sending an ongoing stream of power.

However, when we use those pins that are PWM, they're capable of varying the amount of energy we can see from 0 to 5 Volts.

Certain tasks you're going to attempt and the program will only be completed using these specific pins.

Additionally while we're comparing various boards that are able to use with Arduino, as well as the Arduino microcontroller, then we are going to take a look at how fast the clock is too.

This is an important aspect to consider since it provides us with more information about the speed with which we will use our codes simultaneously.

The higher the speed we can see on this board the more responsive can expect the board to perform as well. Be aware that this will result in you taking more battery and energy.

You must find the perfect balance of power, speed and battery life that is ideal for you.

Another aspect we'll need to look into this is the fact that it will eventually be referred to in the UART. UART.

This will be the portion accountable for determining how many serial communications links the device will to be able to handle on the route.

These lines will be the ones that will come into the system and send information in a sequential way, or as an inline, instead of dispersing them parallel or at the same time.

In order to get this device to work we will need a lesser amount of hardware than other devices, making this device much simpler for us to manage all-around.

For some of the various projects we're hoping to tackle using this board, you might require to connect your Arduino via the Internet.

This is the reason we have to keep a couple of USB drives, and other devices in this gadget. You can connect the device to use it for any of the various online services you'd like to.

There's so much that we can accomplish using the Arduino controller and to understand how to use it, and all the cool things we can do with it is very enjoyable.

We've covered some of the fundamentals to help us begin however, it's time to find out more about the capabilities we are capable of doing with this device for our own requirements.

Chapter 3: Setting Up Our Arduino Board

It is now time to go through and look at a few of the steps we'll need to take to be able to interact using this controller. once the controller is prepared in place and is ready for use, we will be in a position to understand how to perform the essential steps for us to begin working.

The initial step of this is to pick the best board among the numerous options we'd like to use, and let's begin there.

How to Choose Arduino Board Arduino Board to Use

The first step we must take here is to look over our options of boards and choose the one that will best suit our plans.

When looking through some of the possibilities available on the boards There are some aspects to be considered before we make a decision.

One of the aspects that are most important in this case is the amount of power we'd like the board to have based on the programs we would like to run.

It is a good idea to do this if are just starting out and are looking to master some programming, you will need to do so without knowing the specific kind of project we're planning to complete This means that you may not have any idea of the power supply or flash memory that you're planning to utilize.

However, you'll be able to conduct some research and discover that there's a huge distinction between creating a basic nightlight as well as a handful of the other projects that are simple from the beginning to some of the more complex tasks like building our own robot hand.

Knowing the way you're hoping to proceed in all this will be essential to make sure you're equipped with the memory you need to utilize.

If you take a look at all the various aspects you'd like to achieve by preparing this in advance You'll find that it's a lot more straightforward to identify what controller options are likely to be best suited to your requirements prior to making the purchase.

It is important to consider the amount of power you'd like to be able to provide with this board.

After that then it's time to calculate the amount of the digital and analog pins you'd like to include on your board in order to set things for your projects.

Similar to other alternatives, you don't need to be exact in this area however knowing whether your project requires only a couple of pins or an abundance of them can have a significant impact on the board you decide to utilize at all.

If you'd prefer to use this board to master some of the basics of coding , and you intend to stay with simple projects, you might be able to find that having fewer these pins will suffice.

However, if you intend to progress into the board, or are looking to get started and learn some of the more complex aspects and you want be sure to select the board with many more pins in order to get the job completed.

This will need you to think about more thoughts about what you are looking for to be able to finish it.

The next item on the list of things we should take some time to think about is whether we'd prefer to use one of these controllers that is wearable or not.

This will be a decision that is entirely personal depending on the work you're working on and the project itself, but it's something you are able to make when choosing the devices you want to use, and therefore it's important to think about the other aspects.

This is something we have discussed in this guidebook, however it is important to think prior to time whether we'd like to see the Arduino controller connect to the internet , or not.

If you'd like to be able for internet connectivity, you'll have to look for that option when you are choosing the right board.

Be aware that although the majority of these controllers have this feature, it doesn't always work so be sure to think it over prior to time.

The choice of the board you want to use is not as simple as it appears.

It can take a long time and isn't easy due to the sheer number of options to choose from.

The number of pins you'd like to use as well as the power that will be installed behind the device, and much more are relevant depending on what you're trying to find during the process.

How to use the Arduino IDE

After you've had the opportunity to review the various Arduino boards you would like to work with now is the time to find out details about the IDE or the environment, which works with this.

The program we'd like to build or run through this controller must be compatible with the IDE otherwise, it won't function.

That means that we will need to spend the time to download either the desktop IDE to programming, or download an online version the IDE and perform some of the coding within the IDE.

The first thing we could try out using the IDE is to write some of the programs however, we must ensure that we're capable of downloading the program in the format that works most effectively with our desktop computer.

There are several alternatives to pick from, which expands possibilities based on your personal program , as well.

The first is an application for desktops that works perfectly with the Windows operating system.

This allows you to perform the task from your smartphone or tablet together with a laptop or a desktop that is based on Windows to simplify things should you decide to use.

However, it's not an exclusive version of the IDE you're in a position to use and that is one of the advantages of working using the Arduino system instead of any other controller.

It is also possible to find out that this IDE will be compatible with Mac OS X as well should you decide to do this.

It won't work with Apple mobile devices, however be aware of this in mind. However, it works well on desktops and laptops with this software.

Additionally, you're capable of using the IDE for your Linux operating system, too.

It can be used with any of the distributions it, which opens up many opportunities depending on what you're trying to accomplish through the process.

You can choose to use the tool available on the internet to allow you to work directly in the Arduino IDE. Or it comes with 32-bit and 64-bit options in addition.

If you're looking to get going with all this you must ensure that you're doing the right thing and downloading the right version for the desktop-based IDE you intend to utilize.

You are able to start the application is required for installation. go through the various options on offer, and by the time the time is up, you'll be able to get the Arduino IDE up and running for your requirements.

This is an important move to take since it will facilitate the programmer to gain accessibility to IDE and some of the other programs we'd like to use in these kinds of processes.

It is also possible to use it with different sources according to the type of projects you'd like to finish.

How to code your programs using Arduino

The final stage in this procedure is to be able to use our programs and create instructions for the controller we have chosen to use.

We'll look at the codes that we can write using these codes later on However, for the moment we'll take some time to learn more about how to write code specifically for our IDE to ensure that it runs the kind of application we're looking for to accomplish.

Writing code doesn't need to be complicated So don't be afraid to try it even if you're brand new to programming.

There are many different programming languages you could choose to use for this, so it's as simple as you want for it.

The IDE will become a part of our controller, which can make the creation of codes and running them as simple as it is possible. This is the reason we spent time to understand how to utilize it.

After we've been able to practice how to write code and then make sure there's an option to run and execute the code. Afterwards when some issues do come up, we will be able to ensure that we can solve the issue immediately after they pop up.

It is likely that the most effective method to accomplish this is to install the programs you would like to write onto the controller, then check if it runs or not.

It is first necessary for us to join the boards that's the second step here.

How do I Connect with the Arduino Board

When we take a an examination of the options are available on boards you'll see that a majority of them will have a port to connect the use of a USB drive.

In order to begin connecting the board it is necessary to pull the right end of the USB cord and connect it to our laptop computer as well as to the Arduino board that we're using.

This will allow the Arduino board connected to the component of the computer we'd like to use.

If we manage to accomplish this then the software that comes with the IDE that is for our Arduino will immediately begin to recognize the type of Arduino you're trying to make use of.

If this isn't something that occurs then we need to check and select the appropriate one by selecting the dropdown menu displayed on the screen.

Alongside the USB we've talked about, we may also need to pull out, in certain situations in some situations, the TKDI cable or breakout board in order to ensure that our controller will fully compatible with the system that we currently use.

This will include us going through and installing the TKDI into the appropriate board of the controller, like we did using the USB.

In the majority of cases however, the USB will work perfectly for our requirements and we'll be able to depend on it to complete everything.

How do I upload the Arduino Board

To assist us in getting some work completed to upload the sketch and the code already developed, you will need to make the effort to review and select the correct port and the appropriate board you want to upload this to.

It's a straightforward procedure to select the board we would like to use since you only need to take a look at the names of the board and make sure to match it with the ones that are displayed for you.

To ensure that we have selected the correct serial port for this, we are able to pick from a variety of possibilities to complete this task easier, such as those listed below:

If you are looking for the Mac computer, we're going to offer two choices.

If we're working using Leonardo, Mega2560, or Uno We are going to utilize the code below:

/dev/tty.usbmodem241

However, if you're working on board that is Deumilanove or older boards, you'll need the following code:

/deve/tty.usbserial-1B1

For the rest of the possibilities that will be connected using the adapter that connects USB into serial port, will choose the following options:

/dev/tty.USA19QW1b1P1.1

After that, we can begin working using the Windows operating system. We will also be able to see see how it will meet certain of our requirements.

If we're working with the serial board, then we will work using either COM1 or.

If you're working with the board connected to USB USB You will need to use COM4, and COM5 or greater.

It is recommended to look into Your Windows Device Manager to help you determine the port is connected to the particular device or board you are using , and then take it from there.

Also, we can use the Linux system too.

If you intend to connect it with an serial port, then you must apply the following code:

/dev/tty.ACMx

If you wish to make use of the board to create something similar to USB ports, you will need to use the code: USB port, you will have to apply the following code:

/dev/tty.USBx.

Once we've managed to make use of these codes and select the correct port and board for our requirements, it's now time to hit the Upload button before we decide what Sketches is the most suitable from the list that pops up.

A thing to be aware of is that, if you're working on boards that are a little older, you'll discover that uploading sketches is simple however some older boards require going through a reset process or other process for the sketch working effectively.

Running the new Program within Arduino

At this point, we're prepared for the main ceremony.

We must look at the various steps required to run the program on Arduino using the program we created earlier.

The good news is that there are handful of techniques we can use to provide power to the

Arduino board Arduino after the code has been fully programmed on the board.

The first option we'll use is to power this all via the USB connection that we had installed earlier.

Simply connect an electronic controller onto your PC and you're good to go.

The other method is to connect the ethernet cord and connect it in this way by connecting it with the network you require.

The third method we can employ is to simply add the power of a battery and allow it to run as it should.

After you've been successful in getting that power-up and connected in order that you can utilize it and you have the correct input, and you have downloaded the program after which it can be accomplished.

This is where everything will be able to fulfill the task it is supposed to and you are in a position to program and much more using your personal Arduino microcontroller.

Chapter 4: Home Automation The Presence

Sensor And Automation For Irrigation System

Using Arduino

In the initial section of the chapter we'll create a standard automated home system. We will make use of the presence sensor PIR to switch a light on and off. Sensors are connected with an Arduino and generates an alarm signal that indicates the presence of people within a specific distance. When we read the signal from the sensor, we'll be aware of the presence of someone nearby and then turn on a light by activating the relay module.

The Home Automation System: Preference Sensor for Arduino

The motion sensor PIR we will make use of is model dyp me003 produced by the company elecfreaks. The sensor is capable sensing the movement of objects within an area of up to 7 meters, which is a good distance that allows you to turn on the lights and trigger other devices for automation on site.

The process of operating the sensor is very easy. When an object is detected within the range of detection it will trigger an output at a high rate when there isn't any movement within the zone, the output is in a lower position. In summer, the range of detection might be smaller because of the warmer temperatures. The datasheet explains this as well as other details. We strongly recommend you read the datasheet for the component.

Operation of PIR Sensors. Sensors

PIR stands of passive infrared. The primary element is a pyroelectric sensor. This kind of sensor is capable of detecting changes in the radio waves it receives, and emitting an electrical signal that is an output. The sensor is enclosed by transistors in order to amplify the electric signal by radiation variations in the surrounding.

The pyroelectric sensor absorbs infrared radiation using the lens of a fresnel, which splits the sensitive portion in the sensors into segments with lenses that have various characteristics. The sensors that are PIR activated only by certain energy sources, like the heat of animals and humans for instance. The term "passive sensor" is used to be one that does not emit any signals; it simply receives radiation, and is then activated.

Dyp-me003 Specifications

The model dyp -me003 of the elecfreak compact and easy to use. It is easy to use and is extremely adaptable for prototypes and residential assemblies. The sensor module is equipped with an extremely small yellow potentiometer that can modify the duration of waiting for stabilization as well as to alter the sensitiveness. Stabilization time can range from 5 and 200 seconds.

The most fundamental requirements that the sensor must meet are

* Sensitivity and time-adjustable;

* Operating voltage: 4.5 - 20v;

* Output voltage: 3.3v (high) - 0v (low);

* Detectable distance: 3 - 7 m (adjustable);

Time delay: 5 to 200 seconds (standard time five seconds);

* Size: 3.2 x 2.4 x 1.8cm;

* WEIGHT: 7G;

The picture below shows the sensor we are planning to use upside down. The two potentiometers used to adjust sensitivities and stabilization are clearly evident, and are located on each side of the electrolytic capacitor.

The picture below illustrates the primary sensor component. Typically, the sensor is covered by a protective film, such as the image highlighted in this article. In the picture below the film was taken off so that the module could be exposed to the pyroelectric component.

Applications

It is perfect for commercial, home and security automation projects. It is extensively used in:

• Security Products;

* Devices that detect presence;

* Lighting that is automatic;

* Automated activation of buzzers, lamps, as well as commercial and home automation circuits

* Automation and industrial control

Project Description

For our experiment, we're going to utilize our ElecFreak PIR Sensor to trigger the lamp using the relay module. For more information regarding the relay module you can look up all the other referenced within this publication. The project is comprised of the following elements:

"Read the PIR (Passive Infrared) sensor to detect movement within a range of up to 7 meters . Then start a lamp when movement is identified. After a set period of time following when the first detection was made, switch off the lighting."

Hardware Aspects

To put together this project it will require the following elements:

* Sensor DYP-ME003

* Relay module with 4 channels, 5V and 4

* Protoboard;

* Holder and lamp;

* Arduino Uno;

The following assembly shows how the circuit is constructed. The relay module doesn't require transistors for amplifying the Arduino signal that powers the lamp. The relay module has to be connected using either the pin common to one of socket's terminals, and also the NA pin at one of the lamp's terminals. The second terminal of the lamp has to be connected to the opposite terminal on the socket.

Software Aspects

The software is a continuous monitors of PIR sensors in order to determine if the sensor detects any infrared changes (movement of individuals or animations). We declare the variables that find the output pins (PIR sensors output) and output pins (activates the Relay module) as well as in the empty loop () function, we read the output signal from the PIR sensor that is digital, that is , is it high as well as low-level.

#define Relaymodule 7

#define ThesensorPIR 8

```
in ReadingSensor

Setup not in place ()

{

pinMode ( trigger it, OUTPUT );

pinMode (LesensorPIR, INPUT );

Serial . start (9600);

}

void loop ()

{

//

Learn to read the sensor's presence DigitalRead =
ReadingSensor (LesensorPIR);

/ No

if motion detection (ReadingSensor == LOW )

{

digitalWrite (trigger, low );

}
```

// Movement that has been detected. Switch on the light.

and many more

```
{

digitalWrite (trigger it, HIGH );

}

delay (2000);

}
```

The delay is used in the loop function, and based on the time setting you select, it's required to alter this delay.

Put it to Work!

Let's check out how the whole assembly is finished! Below is a diagram of the entire circuit, connected and assembled to power the lamp.

Focusing the mounting onto the PIR sensors

Complete assembly:

Through this article you know the way PIR presence sensors function and how you can incorporate them into your home automation system or other program you'd like. The sensor is easy to install and can be purchased throughout the United States.

In this second and final part in this section, we're going to cover something that is of interest to lots of people: irrigation systems. From small gardens to bigger areas and sites understanding how to automate your irrigation system allows you to be flexible and lets you design specific solutions that meet your needs. In these systems, it's important to determine soil temperature and moisture. To act as an actuator and release water into the pivots it is necessary to have a valve. Of course there is a microprocessor to monitor the sensors and operate the valve based on the soil's moisture and temperatures.

Automatization of Irrigation System The Humidity Sensor as well as the Solenoid Valve

The two major components we will employ for this endeavor are

1. Soil moisture sensor module ;

2. Solenoid valve to allow water into the inlet ;

It is an sensor which measures the humidity of the soil. By using it, you will be aware of when it is important to switch on the irrigation in order to control soil moisture. The solenoid valve acts as an actuator in the system. It uses a signal of 12V to close and open the valve and allow water to flow, or preventing it.

Soil Moisture Sensor Module

The sensor in the soil detects fluctuations in the humidity of the area in which the probe is placed. The sensor is equipped with two outputs: both analog, and digital. This output analog is that of the sensor, and it changes according to the moisture of soil. Digital outputs are the result is generated by an LM393 based comparator which only provides the low and high levels. To achieve higher-precision controls, it is suggested to utilize the analog output so that you can increase or decrease different actuation ranges of your system. With this digital input, it is possible to can only perform an ON/OFF control that is based on two sensor limit readings.

The basic principle of operation is straightforward. The module is composed of two rods and two contacts. It's a sensor whose electrical resistance changes based on the soil's humidity. So, the more wet the soil, the less the sensor's resistance. The more dry, the higher its resistance to the sensor.

Digital output (D0) The fundamental procedure is as following: When there is a low humidity (dry soil) The output is high. If it is not (wet soil) it is at a lower level. A tiny potentiometer in the sensor can be utilized to alter the limit of reference. The circuit used to determine the comparator is located in the photo below:

Ground Sensor Module Descriptions

The specifications for the humidity sensor are:

* Operation Voltage 3.3 5V - V;

* Sensitivity can be adjusted via potentiometer

* Analog and Digital Outputs;

* Led indicator of the voltage (red);

LED indicator to indicate digital output (green);

* Comparator LM393

The dimensions of the PCB are 1.5 by 3 centimeters;

* Probe dimensions: 2 x 6 cm;

Pinout:

* GND;

* VCC;

* A0: Analog output;

* D0 = Digital Output

The output of the analog circuit ranges from 300 mV up to 700 mV when soil is wet, and 700 mV up to 950 mV on soil that has been soaked or with pure water. When the soil is dry, and slightly moist soil it fluctuates between 0 and 300 millivolts. Two outputs from the sensor module were connected to the Arduino to take readings. Based on the readings we can control the valve that is solenoid. We strongly suggest reading the datasheet for the sensor.

Solenoid Valve

The Solenoid valve serves to regulate the water flow. When powered by 12V DC it opens and permits flowing water. If it is de-energized, it closes and shuts off the flow. Its mechanism is based upon the coil (solenoid) that is energized and creates an electromagnetic field which is responsible for causing the valve plunger to move.

Specifications:

* Operating Voltage 12V DC

* Thermoplastic body material;

* Metal elements: galvanized steel,

* Membrane: rubber (standard);

* Entry Opener: 1/4 inch

* Outputs: 3/4 inch;

The valve is compact and simple, perfect for gardening and small irrigation systems. With the sensor for soil it's an the perfect equipment for hobbyists and makers.

Applications

The application of the humidity sensor and valve are typically used in control of water flow systems, like:

* Automatic drip irrigation

* Traditional garden irrigation;

* The irrigation of pots for plants or beds that have plants that are sensitive;

* Control of flow of water;

* Collection of soil information to monitor and analyze;

Hardware Aspects

To build our assembly, we'll require the following parts:

* Sensor module for soil

* Solenoid Valve

* Arduino Uno;

A 12 V AC source external to the

* USB cables and Jumpers;

5.V Relay Module;

This module uses the relay to turn on the solenoid valve as along with the Arduino running at 5V, the relay module does not have enough power to operate the valve. If you're still not sure how to utilize this module visit this blog post about lamps and relays using Arduino. The assembly can be seen below:

Software Aspects

Let's look at how the software is able to detect the sensor and then activate the valve for solenoid.

The pins we define are those that we will read and activating the relay by using three directives at the start of the program. In the void setup function we just initiate the serial port so that we can talk to it and set up the inputs and outputs. The most important part of this is the loop () function. We will take two outputs of the sensor for humidity. Analog outputs are read by the function analogRead (Analog pin) while its digital

64

output can be read using the digitalRead function (Digital pin).

The readings are displayed on the screen, and do a comparison to determine whether we need to close or even open the valve. In the event that the moisture sensor reads reading high this means that soil conditions are not moist, and we need to turn on the relay to allow the valve to open. The relay then gets activated by sending the GND signal. If the output of the sensor is low, this means it is sufficiently moist that it doesn't require to open the valve any more. Therefore, we send an high-level signal to the relay, which will turn off the relay and turn off the valve.

#define pinDigital 13

#define pinAnalog 0

#define pinRelay 7

The float AnalogOutput = 0,

float voltage = 0;

Int ReadingSensor is 0,

The set-up routine is run once every time it is reset

```
Setup is not void () {•

pinMode (pinDigital INPUT );

pinMode (pinRelay, outPUT );

Serial . Start (9600);

}

// The loop routine loops endlessly:

void loop () {

AnalogOutput = analogRead (pinAnalog);

ReadingSensor = digitalRead (pinDigital);

float voltage = AnalogOutput * (5.0/1023.0);

Serial . Println ( "Sainda analog" );

Serial . Println (voltage);

Serial . Println ( "Digital Sainda" );

Serial . Println (ReadingSensor);

If (ReadingSensor = HIGH )
```

digitalWrite (pinRelay digitalWrite (pinRelay);

Other than that

digitalWrite (pinRelay, HIGH);

delay (2000);

}

It is important to note that we attach the valve to our relay with the NA and COMMON pins, which means that we are using typically open contact. If you are using contacts that are normally closed the logic may differ slightly. To learn how to properly use the relay function, read our blog post in which we discuss this feature more clearly.

Working on it!

Complete Assembly:

The output of the serial interface displays the following data:

It is important to note that at this point in the interface in the above it is reading the value 5 mV. This indicates it is dry with almost no moisture. Due to this, it is evident that the output of digital sensors are on an extremely high value, as seen at the top of the page.

With this simple installation it is possible to build the automated system by the help of your hands. It is possible to alter the sizes of the wires to make an ideal system for your garden or plants. It is also possible to use multiple soil sensors to get readings from various locations and create a more precise measurement of the humidity over more of a large area.

Chapter 5: Understanding How Coding Works

Using Arduino

We've spent a bit of time looking at the way the coding supposed to work with the Arduino in the previous chapter, however, we'll explore the coding further to make sure that even the most novice of users are on the right path to accomplishing this for their requirements.

Programming with the Arduino controller implies that we'll have to create and code our programs in one of the numerous programming languages available there.

If this sounds intimidating for you as a newbie however, it's good to know that it's not as difficult than you believe and there are several languages to choose from that were specifically designed for Arduino. Arduino specifically designed to be easy for novices to understand how to utilize.

Similar to how we will see that the study of math has its own set of symbols to represent some of the diverse tasks we are expected to complete such as multiplication and adding We will discover that there are lots of terms and symbols that must be familiar with before we can begin with programming.

If you've spent time using any coding language before and you're able to see that the language that comes with Arduino is easy to master.

Even if you've never had a chance to code before and don't know the right thing to do The Arduino language Arduino isn't difficult and you'll be capable of figuring it all out quite quickly.

Let's begin by learning something about working using coding on Arduino. Arduino microcontroller.

The Structure

The initial part is how to create some of the structures that need to be present in this kind of programming language.

We'll begin with the setting() procedure.

This is the role we will call to when the sketch starts.

It will only be able run once per time following the startup , or after it is reset on the board.

It can be used it for a variety of things for example, starting variables pin modes, as well as for integration in conjunction with libraries.

There are additional terms that can be utilized when we need to add the extra functions we want.

We then need to make use of loop(). Loop functions will need the controller board to cycle over and to repeat the process that you have set up for more than once.

It may continue to run until a specific condition or a certain number of variables you specify is met.

We will define the condition to ensure that it will end the loop, or it will effectively freeze the program you are working on, and you have to switch off the board for it to stop.

The Control Structures

With a lot of the details about the structure in place and in place and ready to go, it's time to go through an in-depth look at certain control systems.

These control systems will be vital to observe as they will provide us with the method by which we will be able to process the input.

Like we can to make a guess on names There will be a myriad of inputs to control which we are able to utilize to decide the way in which we read the data we are analyzing should be.

71

Another aspect that is thought of as the"provisional language.

The language that is used for this provisional purpose is considered during the process of analyzing our data, which means we know what output is required to occur at what moment in time.

Every language that use Arduino must follow this. Let's examine a couple of these.

The first option we'll look at is called the If statement.

This is the very first conditional statement and it will be a simple one and rarely used because of of its shortcomings.

This could be a great method to connect to a particular condition or inputs with the desired output we have set up.

This means that if specific condition we define is met the result or action from our controller will be expected occur.

Let's say we're working with a thermometer.

If we connect this to the Arduino controller, and it is more than 75 degrees in a single time then you can set up the code to inform the board to transmit a signal to the air conditioner, so that it turns off, thus reducing the temperature until it gets back to 75 degree mark and not higher than that at the moment.

If it remains at 75 ° or less and it is not above 75 degrees, nothing is going to occur with the 'if' phrase is used.

It is now time to go to the next conditional statement. It is commonly referred to as the 'if-else' conditional statement.

This will be like the conditional if statement we made in the previous example, however it will describe a different kind of action our controller can perform if the first condition is not fulfilled in the procedure.

This is an excellent method to consider since it ensures that the two possible events will take place according to what's going to occur along the way.

After working on these conditional expressions, it's time to begin to work on some in the loops.

While loops are the first choice that we can employ it is called that of the while loop.

This will be the one to continue to be valid for a long time until the condition you set up turns out to be untrue.

What this will be able to mean for certain codings is that the loop will be performing an action until the parameters you specified aren't being met, then the statement that is going to rule over the assumption you made is not true.

The loop is then completed.

The second loop we can work with is called the 'do while loop'. It is like the 'while' phrase.

You will also notice that it will ensure that the condition to be tested one time prior to the loop being ever tested with the variable, instead of making this happen at the start of the process.

When we're focusing on certain loops that we are focusing on, we must ensure that there is some sort break function appearing.

This will be the point at which you exit the loop. This will make sure that the controller does not become stuck in a loop that is endless while it's doing its job.

Always double-check that you're making this error correctly otherwise, you're likely to be left with an issue with your code.

Another thing worth taking into consideration is what is called"the return function.

The return method is an excellent method to ensure that we can transmit the function we are using and if it does end up giving us something of value after the function ends and we are left in the call function or the function seeking that type of information.

The method we use to do this is dependent on the kind of code we write.

The last aspect we should be looking at will be the "goto" role.

This will be an extremely useful feature that will inform the Arduino controller that it has to change to a different spot, any location which isn't consecutive to the program code.

It will transfer some of the flow into other components of the program. it will be resisted in certain instances by other programmers who use this C language, however, it is a good idea in reducing certain programs we intend to develop using this controller.

The Coding Syntax

It is now time to look over certain syntax can be created in coding with Arduino. Arduino board.

There are some elements of this syntax we can utilize and focus our attention on.

The knowledge of these components will to help us have the greatest results from the process.

The syntax elements which we should pay our attention here comprise:

1. The semicolon

This will be utilized as a kind of punctuation throughout the English language as it assists us finish the sentence that we use in the coding.

We must ensure that the sentence we're concluding with the semicolon is one is considered complete or else we'll find that the code won't work out as it should in the future.

2. Braces with curly:

They will be helpful and perform a number of complicated functions within our code however, the most important thing we must be aware of is that when they're introduced in the beginning it is necessary to complete the process with a pair curly braces to make the code closed.

This will make sure that the information contained in the braces is correct and won't allow anyone other than you to play with the braces.

3. // or one line of text:

If you want to create a reminder for yourself or inform anyone else the way the code is intended to work, then we should make use of this code to start the comments.

It is important to remember when we discuss this topic, we're only discussing only one line at a.

It will not be transferred to the processor of your controller however, it will remain in the code and is a great source for you, and to anyone else doing the manual work of the code:

4. *or a multi-lined comment:

This kind of statement can expand and may extend across several lines.

It will contain only one line of text however it will not contain a multi-lined comment.

You must close the second step of the procedure or the entire of the code after this will be interpreted as an error and won't be applied for you.

5. #define:

This is the element that will make sure that we're able to define a specific variable as having a value which is more consistent.

It will give us with a name that we can use as a shorthand way to express the value.

78

They will not be able to fill any memory space by the chip, therefore it's a good idea to use these chips to reduce the amount of space is available to the compiler.

6. #include:

This will be the sketch that can be used whenever we wish to add a few additional libraries to any sketch we're trying to implement to include additional words or other code language to the drawing, or ones which aren't automatically available in the sketch.

If, for instance, you're going through this process, you might opt to incorporate the C or AVR libraries along with some other tools to make programming easier. This can help in getting the job completed.

a. The most important thing we must keep in mind in this case is that you should not intend adding a semicolon the conclusion of this sentence This same rule applies when working with the #define we spoke about earlier.

If we include the semicolon in this case to end the sentence the code will be able to see an error in this case and the code isn't going to function properly.

The various Types of Data

After we have learned something about the subjects discussed above now is the time for us to take a look at various kinds of data available with the entire.

The types of data we're discussing in this article will refer to various kinds of data that can be handled by the various configurations of programming we're looking to implement.

The data you get via this method is going be passed on to the software you'd like to select to aid in determining the outcome and also.

There are plenty of choices when it comes to the kind of data we'd like to utilize However, some of the most commonly used include:

1. Void:

This will be the process we will employ to inform to the control system that you do not in the position of sending back any data once the function is completed.

2. Boolean:

This will be a type of data that is able to hold two values.

It could be truthful or untrue.

This will be available in conjunction with the functions that you'd like to count on.

The function will return a value that is real or is not true, based on the inputs and the kind of conditions you set in the first place.

3. Char:

This is an instance of a character can be found within your codes, for instance as one single letter.

It could be a numeric character too.

You will often use strings of characters to create the sentences and words that you require.

4. Unsigned character:

This will be similar to the ones we observe in the characters we mentioned earlier, however we will be focusing mostly on the numbers which range between 0 and 255 to identify characters, rather than the signature characters which could be used to contain something negative.

5. Byte:

This will be the type of data we can use which will allow us to keep numbers.

It could also range between 0 and 255, but it is an eight-bit binary system. numbers.

6. Int:

It will be the numbers you can work with and is one of the primary methods you will be able to utilize to store certain numbers you want to use.

7. Word:

When we speak of the term data type, it will keep up to 16 bits of not signed numbers in the Uno as well as on other boards you might like to work with.

If you are working using the component called Due (also known as Zero), you will have to deal using 32-bit numbers using a few terms.

This will constitute one of the principal methods that we have to store the numbers and numbers.

8. Float:

This will constitute one varieties that will be counted as single-digits, and follow by decimal points up to seven before it can be multiplied by 10 up to 38, if you want one that is this large.

It is possible to use this to save some more specific numbers or at the very least, a few which are larger.

They will take on a significant amount of processing power, but they provide 7 decimal places and it's possible that it won't be enough for all your needs.

9. Double:

This will be a type of data which is only relevant when we wish to use the type known by the name of Due that is designed to let us achieve double precision for our floating-point number.

For the various other controllers and boards that work with Arduino it isn't something we'd like to frequently however it could aid us in increasing the accuracy or precision of our operations also.

As we can see from all this, it will be a good opportunity to learn some of the fundamentals we should know regarding the Arduino controller as well as the various languages that will aid us in getting it setup and ready for use.

Knowing the fundamentals and the way we're capable of doing during these situations will help make it easy for us to extract everything from this controller device that we require.

Chapter 6: Turning The Arduino Into A Personal

Machine

When we've been able to grasp a little about the code that needs to be performed by an Arduino-based controller it's time to understand how to put it to use and making it into a personal computer we'd like to make use of.

Although the switches and buttons found on this type of board are sure to be simple to use and offer a wide range of possibilities, there's still a many things can be done using this board, besides switching it off and on.

While Arduino is likely to be referred to as an electronic device, it will assist us in receiving the data we require by using an analog device, so that it can determine a wide range of variables in the world that surround us, for instance, figuring out the temperature or figuring out if the light is up or off.

There are a variety of choices for inputs and programming languages that we would like to utilize to give us the best results, so that the outputs we see are dependent on sensors as well as the code you select.

You will find that many of these sensors and the meaning they are used for can be found at the website for Arduino. Arduino.

To make all this, however it is necessary to make use of the built-in Analog Convertor to Digital built into Arduino.

Then, you will be able to make use of the temperature sensor to determine the temperature on your face.

This will be a device we will be able to identify applying a fluctuating voltage that is present in accordance with what we detect.

To make this easier for ourselves, we're working using a total of three pins that we have on the board.

The first pin will serve as our grounding pin and the second pin will be connected with the source of power we'd like to connect to.

We will need to make sure that the third pin to be the one that will help in transferring this voltage into our control.

This project is coming with a sketch of its own that will ensure that we can understand the sensor's input and, after that, switch on the LED before turning it off and displays to us the degree of warmth that will be displayed.

The temperature setting and sensors will be returned to us in various forms.

For instance for instance, the TMP36 is the one we pay attention to for this particular project since it's going to display the voltage of a different type that could be different from the one that appears in the Celsius measurement, as well.

This will allow us to understand the temperature in a way that we are accustomed for working in.

Remember that this IDE included with this controller is set to include an interface for monitoring serial connections.

This device will enable us to keep track of some of the results from the controller.

If we utilize this monitor serially it will help us to find some of the data which will be related to the condition of the sensors. This will help us know more about the circuit and the program that needs to be executed to make this particular program to function for us.

How Do We Create Our Circuit

With the background set and in place, it's time to look at the first step, which involves making the circuit.

In the next step, we will go through this process and then work on this process in a manual way however, we will continue to calibrate the process to ensure that it works.

We'll utilize the button to aid in providing the reference temperature , or allow the Arduino controller run before the loop is started and use that as our reference point on this, so we will know what to base our work on.

In order to achieve this it is necessary to deal using a couple of the steps below:

1. In the beginning, we must connect the breadboard we're using to the ground.

2. Then connect the cathode on each LED we have to the ground using the aid from the resistor.

Join the anodes of LEDs with pins 2,, using the number 4. They are crucial because they will serve as the project's indicators.

From there, we'd like to be able to attach the TMP26 to the breadboard , by making sure that the rounded side is away from the board you're using.

Then , you must join the flat left side of the pin with the right side of power and then turn the right pin should be connected to the ground.

Then, we can attach the pin into an AO connected to the Arduino we're working with.

The next stage we're looking to work on is the creation of an interface that the sensor can use to facilitate users to interact with it.

You can also work with the cut-out of paper which will look like the human hand, if this aids you in getting this accomplished.

If you're right about it, you'll be in a position to create lips for someone to kiss and then observe how it will appear.

It is also possible to mark the LED's to ensure you understand the meaning from the LED's.

In order to continue this, we'll take out some paper and then cut to allow it to be placed perfectly on top of our breadboard.

We can then design the lips where sensors will be put.

We must ensure that when we perform this, we have some circles we can see using the LEDs.

From there, we're going to cover some cuts of the lips of the breadboard to ensure that the lips continue to remain in the vicinity of the sensor as well as the LEDs inside the holes.

Then , you can apply pressure to this to feel how it feels.

Once we are at this stage, it's going be the right time to look over our code for a moment to ensure that it's going to perform well.

What are some most useful terms?

Constants will be crucial to the subject we're focusing our focus on It won't take much time to realize that they are available with unique names, to make it easier to locate them and utilize correctly.

This is likely to look a lot like variables used in other programming languages however, they aren't easily changeable.

It is necessary to devote some time in assigning an identifier to the analog inputs to make them easier to refer to and, in turn, we can make the instance distinctive and help us keep the temperature we'd like to use as a reference.

Each degree that we discover to be passed on to be part of the temperature reference then the LED is likely to switch itself on.

The temperature will be written down before it will be saved as a floating point number.

This will be the only element of the process in a position to keep the decimal point we require in general.

As we move to the next stage, we will to perform what is called the initialization process of the serial port.

As we explore this particular command, we are going to discover how to use the brand-new command that will be called Serialbegin().

This command will be crucial because it's likely to assist us in setting connecting computers and boards.

This link will assist us in reading the data to get an analog input to the computer's screen.

The reasoning we'll be hearing here is meant to illustrate how fast communication can be that will begin with Arduino.

Then, you can go through the serial monitor within the IDE that comes along with it, making it easier to see all of the information identified, and you can transmit this information to the controller you're working with.

From here we'll ensure that the digital pin is properly initialized and we will toggle it to turn off at the appropriate time.

This is an excellent tool to use as we create the for loop. It is very easy to turn by a couple of pins from the output.

These are the pins on which we will make contacts with the LEDs in the beginning with.

Instead of having to assign unique labels to the pins employing the function called pinMode() which is a function that allows you to utilize them in loops to make it simpler since it's faster to finish the job.

This is a great method to try when you want to make things repeat several times without wasting space. You can also attempt to make them be in sync.

In the course of this process, we'd like to be able to cause the sensor to display at the correct temperature, and then take that temperature reading also.

While the loop is running from above, we need to ensure that we're working with a brand new variable, which we'll name sensorVal to keep the reading from the sensor.

If you'd like to take the time to look at the sensor you only must go through and call the function of the analogRead() that will then be able take a single argument.

Remember that this isn't the only thing we have to focus on in this area.

It is also important to spend time going over and then transfer the data of the sensor to the computer.

The function we can see in Serial.print() will be able serve to help us to gather all the information from the sensors on the Arduino controller and transfer the data to the computer you prefer.

You can then utilize your serial monitor to assist you look up any information you'd like to know.

If you've taken the time to complete these steps and added the parameter Serial.print() in quotation marks, it's likely to be something that is visible in the text you wrote.

Furthermore, if decide to use the variable as your parameters for this, then it will show on the screen as the values of your variable from the beginning.

It's time to work to the next step and transform sensor's reading into the form of a voltage.

With a little math knowledge is enough to look over and figure out the correct pin voltage you'd like to use for your.

The voltage could range between 0 and 5 volts. It also has some parts that will provide us with.

It is crucial to keep in mind that you'll have to create the float variable to have it stored in the location you want it to be.

From here, it's time to cycle through and adjust our voltage up to a satisfactory temperature before it is loaded on our computer.

The data sheet we will see using this sensor is likely contain a lot of data which is comparable to what we're going to find when we discuss the voltage at which the output is.

The data sheets are merely going to serve as our electronic guides.

They are usually developed by engineers, and are created in a way which aids other engineers who are involved in the process.

Based on the information we'll receive from the sensors, every ten millivolt will correspond to about one degree of variation in Celsius.

Additionally, we will then be able to look through and determine the values we would like to be the offset to the values that are lower than the point at which we can freeze.

In other words, if we're working with minus 05 of the voltage, and you can multiply it by 100, then you'll be able to determine the temperature that we are looking for our work.

In the event that we are performing this procedure and discover that we have less heat, it is possible to arrange this so that the lights shut off.

If you are working using the temperature of start that we have chosen, it is possible to create the 'if-else' clause to let the program decide when to switch on the light and when to switch off the light.

Utilizing this as the temperature of reference and as a good base to work from it is possible to turn on the LEDs when you notice a difference of only 2 degrees.

The program is then installed to search for a variety of values while we go at the various scales of our temperature.

After this after that, the next step of the procedure we'd like to do is switching on the LED to lower the temperature.

"&" operator && operator is going to be a symbol for "and" in a more literal sense, and could assist us in this.

This process will operate in a way that allows us to see whether there are any other circumstances that are triggering this procedure at all.

It is then possible to switch things up and go through an approach that allows us to create a medium temperature that will switch on the two LEDs whenever we would like them to.

When the sensor detects it has sunk down a little or is between two and four degrees below or above the baseline , based on what we're trying to accomplish, the block of code going to switch on the LED. This will be visible in the pin 3.

Of course, this is only one of many examples we can employ in working with a range of coding can be done using Arduino. Arduino language.

When we put all this along with some fundamentals of coding and the knowledge included inside the C language that we'll talk about in a bit and then include a little more function to what we're seeking on this board.

Chapter 7: The C Language

Now that we've been able to take an in-depth look at what the Arduino language will look, we're going to take it to the next level and examine deeper into how to use the C language.

This is among the most effective options for working with the Arduino board. It is worthwhile to get this board and master the basics of it.

As you work on this language, there is likely to have components that you are able to work with that could be somewhat confusing and may will not make much meaning until you start your program. That will take a little difficult to master when compared in other languages.

However, the devices that are likely to perform very well with this type of controller, including the LEDs and sensors will depend on the particular outputs and inputs we will be working with here.

Many of the programming languages that we find on the internet and more be used to complete the tasks we would like to accomplish with

Arduino. Arduino board, however you'll find most of them using C. C language is usually regarded as the most effective one, and it is the one we'll concentrate on here.

The primary system that we use for any type of technology we intend to develop will consist of the control device. It will be referring in the direction of the CPU, or the real microcontroller we spent the time discussing before.

There will also be some minor variations that will be apparent when choosing the controller we want to work with.

But, remember that these controllers aren't capable of being as efficient to the standard microprocessor.

It will contain the inputand output ports, as well as all the hardware functions we need.

It is evident that the microprocessors we're talking about will be linked to external memory.

Typically, controllers will contain enough memory.

But, when we refer to this, we're necessarily talking about larger sizes.

There is a possibility for the Arduino controller that we're using with Arduino to only have one hundred bytes or less of memory for the basic applications it runs.

The register will be the sole place in which we will be able to conduct mathematical operations that are logical that we'd like to perform.

In the case of example, if we want to execute an example of adding two variables, then the value of the variables is something will be required to transfer into the register.

There are many advantages of working using the C language It is essential that we understand how to use it correctly.

It's similar to the majority of work we've completed using the Arduino language as well, and we'll take a take a look at some of the options we can accomplish with it in this guidebook , as well.

Before we get into that but, we should review other fundamentals that are part of this language. In particular, we will explore the concept commonly referred to as memory maps, and how this will aid C make the most of the Arduino boards we have.

Memory Maps Memory Maps

Each memory byte will be visible within our computers will be with its own unique address that is connected to them.

If this address isn't present, the procedure will be a mess because it's lost the ability to determine the memory we're hoping to connect with.

To make it easy You will find an address in memory that will tend to start with zero , and will rise and climb in memory as it progresses even though there is the possibility for a specific location to become more particular or adhere to the rules that are more distinctive.

It's also possible that in certain cases that the address you're looking for might not correspond to the output port or input port when it's time to use external communication.

Most of the time when we're working with this type of language and other forms of communications you'd like to manage in this area, it's going to be necessary for us to look through and create an image of memory to determine the location and state of memory on our system.

This will be a huge undertaking and it will result in a huge assortment of memory slots that we will have to manage during the course of.

Anyone who is exploring these maps understand that it could take time. They also need to at minimum spend a part hours working on addresses with the lowest valued position at the top , while other people who create the map and assign the final address at the bottom of the map will be as well.

It is evident that each of the locations of the address will indicate the location where we're in a position to store additional bytes within the RAM of our personal computer.

When we read the subsequent chapters in this book, you'll be able to see that we're going to spend a lot of time studying some of the fundamentals that comprise C. C language.

There are a variety of languages that are compatible together with the Arduino board, however none of them is going to give us the power and the functions of this device like you will find using C. C language.

If you already have one of these languages, you are welcome to incorporate them to the mix and utilize these languages instead.

If, however, you're an absolute beginner in programming and programming in general, then we'll spend an additional time with this book, focusing on the fundamentals in the C language and all the exciting things you will be able achieve using it.

A few of the fundamentals we'll cover are specifically for the Arduino board and the features we can accomplish with it.

This can allow you begin to use this microcontroller to control certain of your own projects in the beginning.

Chapter 8: The Logic Statements

We have now spent some time working on the C language, and at the very least consider why it's an ideal language to work with and explore in relation to this board, it's time to look at some of the logic assertions in the language and the way they will assist us while we do certain tasks on this sort of project.

The first circuit we had a chance to discuss before was designed to be fairly simple, as it was designed to make making these circuits as simple as it could be.

It is the right the right time to think through and alter things.

If it is the right the right time to begin working on programming using our controller, where the input is going to have an impact on the output, it's the time to start working on the process of coding, also called logic statements.

The logic statement is essential in the coding process, regardless of regardless of whether we're using C or not. C programming language, or

107

not, because they're among the best methods for the programmer to ensure whether this variable's value going to be evaluated against the other values available.

The other value is likely depend on the things we're hoping to measure with our own code however, it's likely to be an unchanging object and it may also be an variable.

If we make use of these kinds of logical assertions, it's likely become one of the strategies employed by the programmer to get the most control over what's going to occur in their designs but without the programmer being able to predict the outcome at any given the time.

In order to help us follow the steps of creating our own logic assertions, we must first enter to the IDE that runs on the Arduino and then follow an exact path.

The most effective path in this instance is File Examples 02. Digital, Button.

Please take a moment to note how similar the code we have just written or the route, is to other codes we've created in this guidebook to date.

It is evident that it is the same pattern used by these controllers, which is likely make them more user-friendly in the future.

We must examine certain variables, and the one that's most important to us now will be the pin of the button while the second one will be more related to the status that the button is in.

This will tell us whether or not the button Is active.

While in the configuration mode, we're working with the pinMode in order to ensure that the pins are set up when we get there.

However, for the scenario that we're in the chip will have the button pin that has the direction INPUT that will inform to the processor that current is supposed to go into the direction of INPUT, not moving out.

When we begin to work with this type of loop, we're likely to be able to start the program, and the first line will be able to add another feature to ensure that things get completed.

In this case, we'll be able to use the code that comes with digitalRead() which will be the equivalent that goes with the commands of digitalWrite() which we discussed earlier.

It is also a great source to refer to for information on the way these Boolean codes are intended to function.

This will refer to the logic of statements which is why it's time to dig more in depth with these.

The result we can observe from this study is dependent on whether there are any specific kinds of conditions satisfied.

The command is expected to be awesome since it is able to pass through the ideal conditions for this section of the code. it is something we'll hear about in the coming months of time.

While we're here we should also examine the state of the button and whether it has been press.

After the button is press, it is going to reveal the results of the HIGH.

It is an expression that can form an in this case, and it will be the part of the code that is going to appear and be incorporated into our curly braces code.

This will allow us to judge whether the system will behave in according to the way we prefer.

We are about to look at an alternative statement that comes with some curly braces and more.

Else is a reference to the fact that if an input is passed into the statement, and it isn't able to connect with the previous statement then it's going to move to the section of the code within the curly braces, and work with the curly braces instead.

In the example we're working on here it is going to be working using the electronicWrite() component of the program to signal the chip inside the Arduino board that it's time to switch off or turn of , similar to what we did previously.

One thing we have to keep in mind when doing this is that even though the sketch will include the else clause, it's not necessary to include this in all sketches we draw And sometimes, we are able to use the if-statement without needing to utilize the 'if-else' statements.

It is also possible to configure it in a way that if a specific requirement is not met the program will not proceed to run the next sequence of instructions. In certain programs this can work well.

With this knowledge in mind, it's time to take a more closely at ways to use our controller to make it blink the LED only the moment we press some of the buttons.

There are several simple circuits to select from this controller, but it is necessary to spend some time to determine how to operate with each of them.

To assist us in completing this endeavor and determine what is required to be done to complete it, we'll need to be able to talk about these areas:

File   Examples   0f.Control
WhileStatementConditional

The very first element we'll be able to take from this sketch is one that is pretty similar to the task you're doing.

We will examine all the variables will be required before establishing them, before setting pins to make adjustments to the settings which will be either the input or output.

After we've completed the loop that is included in the code, we're likely to gain some familiarity with how our while statement will appear like. Some of the codes is required to use is:

```
while (digitalRead(buttonPin) == HIGH( {

calibrate();
```

This is going to give us lots of details even though it's not huge and doesn't seem like there's a lot of information of information.

Because it's so tiny, we must actually split it in such a way that we can observe the contents inside.

The first thing you should take a look at will be the while assertions.

They are crucial when we incorporate something to the curly braces to ensure we can establish our requirements.

This leads us to the problem which will be present at the very beginning of the code.

When the button has been activated and we call this our condition, then we must examine the pin that ought to be linked to the button to determine whether it's pressing or not.

If that's the case it will be in a position to calibrate the function, which is one that the user is able to explore and define later.

What this code will mean in this case is that when the program detects that you've got an instruction to follow and calibrate, it's going to go back to the instruction you set up for that task.

It will then execute the commands before returning to the previous portion of code.

It is also possible to spend an hour or so studying this specific function because that is what it is being referred to at present.

We'll examine the code for void calibrate() part of|section of|the code.} the code.

In the beginning it's going to appear familiar since it's identical in the way that loop() and set() functions we had previously used.

What this code will be telling your compiler that you want to to define a function that has names like calibration and it will then return the empty string to you, which does not show any reason.

This is the point in which we will look for the meaning behind all this signifies.

We are going to be taking the time to understand that the definition of the calibration implies that if we follow the same procedure and enter the exact word into our code into a different section, the compiler will look for a function from the exact same source and will continue to perform in a similar fashion similar to the way we're doing it right now.

What happens when we receive a return of empty space?

We've never had a chance to discuss the possibility of returning to the void, since the goal of this article is to suppose that the loop will continue to function and return to us something.

This particular function is effective however it isn't always true when we put in the void.

If the function is able to go through and finish any of the directions that are within our braces, it will give us a number for the operation.

It can come in a variety of varieties, like the void or no return, however it's possible it is an integer, or another kind of number that is compatible with the calculations we need to do.

To simplify the process and also to find out what we can do now, let's say we're working on an operation.

Instead of sifting through and working out the earnings per week of employees employed by company A then we'll change it up and implement the capability to deal with floating numbers that will include all the figures of the earnings, allowing us to apply them to a new location within our program.

If this is the way we'd like to do Which task should we select?

This is a great method to think about how we plan to utilize our functions and how we can use the logical arguments to help us with our specific tasks.

As we're able to imagine there is a vast range of logic-based expressions that we can use in the Arduino language and all will have various purposes.

It may be difficult initially however, they're going to make a huge impact on the results you can achieve from your board. They will make sure that the controller is capable of working in a way that is based on the input and outputas well as the various other conditions that you set in the process.

Chapter 9: The Current And Energy Meter And

Audio Module Wtv020-Sd, With Arduino

In the beginning that follows, we'll build an AC power meters (alternating current) by using Arduino UNO. Arduino UNO. Energy meters are a part of everyday life in every home, building or business. A variety of models of traditional companies like Clamper or EDMI are in the market, with different prices.

The key components of the meter are the current and voltage sensors, and the technology/measurement specifications will dictate the final project price; we will use a 100A non-invasive AC sensor SCT-013.

Energy and Current Meter using Arduino Current Sensor

The energy meter consists of a voltage and a current sensor, which is linked to an electronic circuit which in this instance can be identified as that of the Arduino UNO. The two primary electrical variables, and from them, we can determine the current electrical power as well as the amount of energy used in the time period and this is determined by the instantaneous power that is integrated into the measurement time.

If the power of the electric source is constant, you can simply increase the amount by time of consumption. For example, a lamp that is 100W running for two hours consumes 200Wh (watt-hour). If a function produces the power of electricity, for example p (t) The energy is calculated by using the integral of the p (t) within the measurement time.

In the scenario we will be looking at it, we'll measure the instantaneous power of the sensor currently in use. The energy is calculated by using an equation (which is exactly the function of an integral). Let's take a look at an integration time of one second. Each step will determine the electrical power, then find the energy that corresponds to this interval. We will then combine the values to an undefined variable that will be displayed in the LCD.

In an LCD the current values and energy that we calculate at the time an Arduino gets powered.

The Current Sensor 100A is a SCT-013

The current sensor SCT-013-000 can measure AC voltages (alternating current) up to 100 A RMS. The sensor is not invasive device and, therefore, is not connected to the measured circuit. In actual use, it's the equivalent of a current transformer having the ratio 100: 0.05 (in the instance of the sensor used in this instance, you should verify the spin ratio for your own sample) This means that an initial current of 100 A is displayed to the second as 50 mA. This sensor is not designed for DC directly current measurements, but only for AC.

Remember that whenever we talk about current we're talking about RMS values. In other words the current that is recorded by the device is higher that 142 A that is that RMS number multiplied 1.4142 (square root of 2). Refer to Your AC circuit book for a refresher on the concepts in question, if needed.

The sensor is equipped with an opening that the wire carrying the current that is to be measured is required to pass (in the case of a home the wire could be the neutral or the phase). This wire is the primary transformer, which generates an equal amount of current in the secondary in accordance with the rate of transformation that the circuit.

The main differentiator is that this device uses a current output and it is not a voltage signal. This means that an additional circuit is required at the output point to transform the current output to a voltage which can be read out by the Arduino (we'll go into more detail about this later).

Specifications

* Current input 1 - 100A;

* Current output: 0-50mA

* Ferrite, the primary material.

* Opening dimension: 13 x 13mm;

* Working temperature: -25 deg C to +70 deg C;

The website of the open energy monitor includes a complete description of what CTs (Current Transformers) sensors are made. Visit it here! (https://learn.openenergymonitor.org)

Applications

Energy meters are utilized in every project that requires them to track the consumption of electricity of the circuit or installation. Current sensor SCT-013 could be utilized, for instance:

* Home automation projects for small homes;

* Design and prototyping for energy meters.

* Protection systems to protect against voltage and current surges

* Lighting systems;

• Energy Efficiency Studies;

Description of the Project

This project is comprised of the following elements:

Make use of a current sensor, and show on the computer screen the results of the energy and current consumed as of that measurement circuit turned on up to the point at which it turned off.

Hardware Aspects

A different circuit is used in order to transform the power into the voltage level that is detected by Arduino is described in this article. In essence we have to determine what the relationship is to the input as well as output. To do this, we simply divide the current of the input by the conversion ratio. In our instance, 100/0.05 = 2000. Therefore, we can tell that the current measured will be the current output multiplied by 2000.

In order to be able to detect the output signal, we must transform the output current into an appropriate voltage that is within the range of measurement for the Arduino. To do this, we utilize a load resistance in combination with the circuit. In this example, the current is alternating, this means that it is a positive cycle and negative. Because the Arduino doesn't detect negative voltages it is also necessary to make sure that the voltage remains always positive.

Sizing Load Resistor

In order to calculate the load resistance and determine the voltage that it can handle must be 2.5V (maximum voltage that can be read from Arduino divided by two). Arduino divided into two). To make sure that the voltage on the Arduino pin varies from 0 to 5V and above, we'll use the smaller voltage divider to increase 2.5V on top of the current voltage at the source of our load resistance. This will result in an average voltage of 2.5V.

It is also important to know the maximum amount of current that the load resistor Is likely to be exposed. Assuming that the highest RMS output voltage is 0.05A The peak current will be 0.0707 A. Therefore the load resistor we will apply is:

Load = Vmaxsensor/Imaxsensor = 2.5V/0.0707A = 35.4O

The most similar to a commercially acceptable value would be 33O. This is our resistance to load.

At the peak that are present, the power that the resistor will dissipate is 2.5 + 0.0707 equals 0.177 W. Understanding this figure is crucial to choose an appropriate resistor robust enough to handle the power that is released. If these values are met that the voltage of the resistor can vary between 2.33V at the peak of positive voltage (33 * 0.0707) and -2.33V at the peak of negative voltage (33 * -0.05). Also, the voltage of the Arduino pin will fluctuate between 2.5-2.33 which is 0.17V up to 2.5 + 2.33 = 4.83V. We assure that the sinusoidal current will be converted into an sinusoidal voltage which varies within the range that is measured by the Arduino.

Each of these values is recommended from an energy monitor that is open that has created an app which calculates the measured current value from the voltage signal we have just constructed.

We are still looking for the calibration value that is then passed in the form of an argument emon1. currently running function (pin the function, calibration) which we're going to employ. The calculation of this parameter is in the following manner:

Calibration_Value = 2000/33 = 61

In the case of 2000, it is the ratio of transformation (from 100 A to 50mA, where the current is multiplied by 2000) as well as 33 represents the load resistance determined just above. This means that we are in a position to utilize the function within our program.

Be aware that this is an "huge sensor" with the capability of measuring as much as 100A. The current output is extremely small, only 50 milliamps. This means that lower amounts of current, like 1 A such as caused extremely low voltage variations on the Arduino port. If you have an A-level current, for example, the voltage in the secondary of CT would be equal to 50mA divided by, which is 0.5mA. For the load resistance this current causes an increase in voltage of 16.5mV that is very difficult to gauge and highly vulnerable to noise and circuit. Therefore, this model SCT-103 100A is particularly suitable for loads with larger capacities.

If you're looking to verify the bench design of a small power source, like we recommend you use an SCT with a distinct transformation ratio. For instance, SCT-013-020 that transforms currents of up to 20A into voltages of 1V.

Important Tips:

* The SCT-030 100A sensors are equipped with a built-in burden resistor. I tested the value of 38 Ohms. Take note of what you can get from your device too. When you've got an existing Burden resistor, it's not required to put the 33-ohm resistor to the circuit. However, in this case you must change the value of calibration (I_calibration) is to be changed to 30.

This is because the reference voltage for Arduino ADC converter Arduino ADC converter should be 5V. Always use an external source of power to supply power to your Arduino. For my Arduino Uno I have the AREF voltage is 4.98V. If you are using USB to supply power to your Arduino it will be able to measure the AREF voltage should be under 5V (4.7 to 4.8V) Therefore, the voltage measurements are wrong.

The circuit in question is extremely susceptible to interference from electrical sources. Make sure you are away from the circuit. If you are able you can use cables with shields for connecting the circuit with the Arduino.

Sensors SCT-013 were created to measure currents of up to 100 A. If you require smaller currents, it is possible to run your AC circuit wire for more than one turn over the center. This way the current measurement is multiplied with the amount turns. Check and then adjust to adjust the calibrator factor.

List of Components

The most important elements for the project include:

* Arduino Uno;

The current sensor is 100A. SCT-013

* 33R resistor, and at the very least 1/4 W

* 2 100K resistors;

One capacitor with a 10uF value

Circuit

A third component that can be added to the circuit could be an amplifier that acts as a voltage source that is connected to the voltage divider and load resistance. When using high-value resistors the load effect is reduced however, in order to eliminate the connection between both circuits (voltage divider and load resistor) it is suggested to incorporate an opamp into the voltage follower configuration , or another optocoupler.

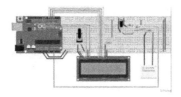

After making these reservations, the circuit follows following:

Software Aspects

We will be using the energymonitor library that was developed.

#include "EmonLib.h"

#include < LiquidCrystal .h>

Definition #define Vrede #define Vrede denotes Vrede 129 as the name of the network RMS voltage (value measured by a multimeter)

"#define" InputSCT 2.denotes the analog channel of the signal SCT. Pin A2

#define LCD_RS 9 * Definition of the LCD pins

#define LCD_E 8

#define LCD_D4 3

#define LCD_D5 4

#define LCD_D6 5

#define LCD_D7 6

#define I_calibration - 60 // current calibration factor – read my advice in the tutorial

EnergyMonitor EnerMonitor Creates an instance of EnergyMonitor

LiquidCrystal lcd class (LCD_RS, LCD_E, LCD_D4, LCD_D5, LCD_D6, LCD_D7);

Double Irms equals 0.

Time = 0 for floats.

Setup not in place ()

```
{

lcd. start (16 2, 16) • Configure LCD 16 columns
two lines

lcd. clear () Clear LCD screen

EnerMonitor.current    (InputSCT    I_calibration)
/configure SCT pins and the calibration factor

lcd. setCursor (0,0) / select row 0, column 0.

lcd. Print ( "Irms (A) is" ) // display text

lcd. setCursor (0.1) // choose row 1, column 0.

lcd. printing ( "Ener (Wh) is" ) // display text

}

void loop ()

{

Irms = EnerMonitor.calcIrms (1480); // RMS
current calculation

lcd. setCursor (9.0) // choose row 9, column 9
```

lcd. print (Irms) // displays the current value

lcd. setCursor (9.1) • select row 1, column 9,

lcd. Print (Irms * Vrede (Tempo/3600)) (Irms * Vrede * (Tempo/3600)); calculation of Watt/hour

Time +;increase Time

delay (1000); // 1 second delay

}

Within the codes, it's worth noting the following aspects.

Its functions are a part of EmonLib.h library are used by an object of the type EnergyMonitor;

* To measure current we need to begin to initialize the object with an operation: EnerMonitor.current (pin or calibration);

To determine the current value value, we invoke our method EnerMonitor.calcIrms (1480) where 1480 is the sample count that are used to calculate the current value.

* To calculate the energy, divide the current with the voltage of grid (set by the Vrede variable) and

also by the duration. Since the loop is delayed by a period of one second, on each repetition, we divide the power derived (Irms Vrede) by 1/3600, and calculate the amount in watt-hours (1 second equals one hour multiplied by three times 3600);

In the rest in the program, we set the pins and LCD to read the data and display the information.

Put it to Work!

Note That the wire to be connected to the sensor is just one, and we in the least, we should not pass the two wires connected to the load supply to the sensor because, with the current flowing in opposing directions through each wire, the magnetic fields that are associated with the wires are cancelled and do are not sufficient to stimulate the core magnetic that is the CT.

Final assembly:

It is important to note that for this measurement of energy the measurement was the measurement manually in order to gauge how much voltage is present in the electric network using an instrument and then use the measurement in the software to calculate a number. Another step to enhance your program is to incorporate an additional circuit that can determine the voltage across the bus. Instead having a variable which has to be constantly checked manually the system can already monitor the voltage of the bus and display the energy measurement based upon it.

In the third and final section of this chapter we'll discover how to utilize this audio device called WTV020-SD. It allows you to integrate audio components into your Arduino projects like music, voice messages and custom alarms, and other. In this short overview to the module, you'll be able to utilize the WTV020-SD device efficiently and easily for your project.

Audio Module WTV020-SD incorporating Audio Elements using Arduino

The WTV020-SD is an microcontroller circuit designed that allows voice recording. It comes with the SD card slot that has capacities of up to 1GB of memory, which provides a lot of flexibility when working on audio files in projects that use Arduino.

Through the module that is available, you can utilize files that are in WAV as well as AD4 (compressed audio) formats which conform to FAT format. WAV format files have an average sample rate of 6KHz up to 16Hz. The files that are in the AD4 format come with a sampling frequency of 6KHz up to 36KHz. For more details on AD4 format, click here. AD4 format, refer to this page. (https://www.filedesc.com/en/file/ad4)

Below is an image of the module.

The module can operate in a number of modes. Are they:

• Audio Control Mode: Stop/Play Next and Volume +- functions

A One-To-One Control Mode switch: switch is activated by the voice that activates three voices and also adjusts the volume;

* Control Mode Power Loop: Once the power loop is on, there's no requirement to turn on the output or input, and the recording is directed entirely into the SD card.

Second-Line Control Mode It is free to play any file from any location. Reproduction in combination can be utilized in this mode.

The nominal voltage of the module will range from 2.5V to 3.5V to 3.5V, which means we can make use of the 3.3V channel within the Arduino to charge it. The ADC converter that is used by the module has 16 bits.

Connecting the microcontroller (Arduino) directly to the module, we could make use of four pins for communication in order to control the function of reproduction of sound. This is called the Second-Line Control mode.

For Audio Mode, push-buttons are utilized to turn on the module's pins and carry out the same functions for playing back music previously mentioned. The two other modes are not used as much due to less frequent situations.

Applications

The WTV020-SD module can be utilized in a variety of applications, including:

* Car alarms;

* Voice messages;

* Music player

* Parking Sensor

* Medical and smart home devices.

Any project in which the integration of audio elements is essential is a good candidate for the WTV020-SD Module.

Project Description

The sample project includes the following elements:

Make the files and store them to the SD card of the WTV020-SD device, the first with WAV format, the other one in Ad4 show a menu on the serial interface that allows you to select which one to play and the length of time.

Hardware Aspects

The list of elements for the proposed project are as the following:

* WTV020-SD module;

* Arduino UNO;

* Protoboard;

* Speaker to Arduino;

A short list of parts available on the market.

The pinouts for the WTV020-SD device can be shown on the following table. It is taken from the datasheet that was originally published. In this project, we'll make use of those power connections (3.3V as well as GND) along with four additional communications pins.

Pinout WTV020-SD. Source: Datasheet:

Then, we'll be able to see this connection:

Pin Module WTV020-SD Arduino UNO pin

RST (pin 1) Digital 2

P06 (pin 15/BUSY) Digital 5

P04 (pin 7/Clock) Digital 3

P05 (pin 10/Din) Digital 4

Make the circuit as per the following diagram:

Software Aspects

In order to control this module we'll utilize for control, the WTV020sd library. The library was released through the Arduino Community. Prior to recording and then running the audio files, it's essential for you to reformat the SD card using

FAT file system. You can also save a file using the AD4 format as well as one with the Wav format.

To convert WAV files There are many applications available online (ex1). To convert to AD4, we recommend this article (https://www.buildcircuit.com/how-to-convert-mp3-and-wav-files-to-ad4-format-wtv020sd-tutorial/) that presents several methods. The source link for the library's publication contains an example of how you can make use of the different functions offered by the module. The program looks like this:

```
/*

Example How to control a WTV020-SD-16P Module to play voice recordings using the Arduino board.

Created by XYZ on August 16, 2019.

In the public domain.

*/

#include
```

2, The pin number for the reset pin.

Integer clockPin is 3. / The pin number for the clock pin.

Int DataPin is 4; The pin number for the data pin.

Int busyPin is 5 / The pin number for the pin that is busy.

/*

Make an instance from the class Wtv020sd16p.

1st Parameter 1. Reset pin numbers.

2. Second parameter 2nd parameter: Clock pin number.

3rd parameter 3rd parameter: Data pin number.

4th parameter 4th parameter: Busy pin number.

Chapter 10: The Api Functions Api Functions

If you are ready to use your Arduino, you'll have to look at the API functions, as well as all of the amazing details and functions that come with them.

The API that is included in this controller sure to be extremely rich in capabilities, and the options programmers can take advantage of will make it much simpler to work with the device.

The team responsible for Arduino's development Arduino machine has performed a fantastic job using the API and ensured that it's completely functional and gives the programmer various options regarding what they can create using the Arduino controller to develop their own projects.

We'll take a few minutes in this chapter, taking a look at the ways we can use these tools in greater in depth, so that we will understand the functions and the capabilities, and discover how beneficial this could be to our work.

It is the Digital Input and the Output

The first item on the list of things to do is to work with the output and input digitally.

There will be many features that we could develop using Arduino API. Arduino API, which is likely to allow us to communicate and work using the digital pins which are available.

We must look at the three kinds of these, so that we can put our board to perform.

To begin in this direction we must follow these codes:

pinMode(pin INPUT - OUTPUT INPUT_PULLUP)

This will be helpful as it lets us identify a particular pin, and then specify if the pin will serve as an input or to be an output.

The latest models of this Arduino will to be able their pins activated by pullup resistors. That's the reason we added this feature to it.

From this point, we are in a position to examine the input and output of Analog.

Alongside using digital pins in many of the projects we work on, it is possible to add some analog pins, too.

The analog pins are likely to be beneficial because they'll help us detect the voltage that other pins will give us.

The readings should range between 5 and 0 volts.

This upper range likely towards 1023 in this case We will observe the low range will move closer to zero.

The code we have to implement to ensure that we're reading the correct pins are analogRead(pin).

This will help us understand the voltage on any pin we choose to study.

We will then find an integer between 0 and 1023 as we mentioned in the previous.

Or we are able to work with a different code, the analogReference(type), which will then help us to configure this over to voltages instead of giving us the reference point that we need.

The method we select to analyze the numbers may depend on the type of board we use.

Then it's time to explore some of the more sophisticated options that are available through the input and output to the Arduino board.

They are unlikely to fit in with either of the categories we've talked about since they are not digital or analog.

However, they'll become more sophisticated categories in this regard and we will be able to create a lot more using the boards once we incorporate them into the mix.

To begin, we have an initial tone(pint frequency, frequency, or Duration, and frequency).

This will allow us to can specify the frequency we are looking for and create a square wave with the frequency on the pin that we're using.

We can then work with noTone(pin) which will assist us to stop the tone generated by the tone function at times we want to.

There is the option of adding a different kind of pulse too.

We can use the algorithm of pulseIn(pin value) which will enable us to detect the pulse on any pin.

If the pin is oscillating between low and high at the moment, it will return the microseconds that separated the low and high.

Since the pulses might not be perfectly even at the moment it is possible to go through and select whether you would like to see the HIGH pulse as well as the low pulse, based on the one that works best for your particular task.

Working using Time

We should make a detour here and take a look at some of the basic concepts of working on time.

This function is compatible with our board and will allow us to with any project that is time-sensitive, for instance, making it our controller's scope.

The first method can be introduced to this code is called"the delay(value).

We've seen it a couple of times already and the reason behind making use of the technique is it permits us to put the sketch down that we're working on for a specific duration.

It is possible to add in the value in integers of how long we would like the sketch to stand still be aware that this is by calculating milliseconds.

Then we are able to work with a second code that is going to be delayMicrosecond(value).

This is an excellent program to play with, but it does not have the same functions we had in delays() function we used prior to.

This will be done in microseconds instead of milliseconds. If you require a different frame to work with here then this is the time frame you should select.

Working with Math

In the case of programming, there's going to be plenty of math.

The math functions are available in the Arduino API will be like the math library included in the C language, however, they are usually viewed as being more user-friendly.

This is due to the fact that you are capable of using these libraries without the need to actually transfer the math libraries you require.

Even if you are working on programs that don't require an excessive amount of maths at all You can still gain from these as you never know when it might be introduced.

In the majority of cases the math you'll work with is straightforward and will incorporate a lot of the various assignment operators we spoke about in the past.

It could be options such as multiplication, subtraction, division and division to mention the many options you can employ.

What Characters Are They?

The next thing we must look at is the characters.

While they aren't common in this type of language but it's still vital for us to understand these words whenever we have a job which is geared to us.

The most well-known characters we're capable of working with include:

1. isAlpha(character):

This one will be returned regardless of whether the character we're using is an alphabetical character.

2. islphaNumeric:

This method will return a character regardless of whether the character is either alphabetic or numeric.

3. isAscii():

This will reveal whether or whether it is one of the ASCII character.

4. isControl:

This will determine what the status of this character is. can be considered to be a control-character.

5. isDigit():

This will tell you whether the character is an integer.

6. isGraph():

This will determine whether the character contains visual information.

Space for instance, will not have any of this data visual.

7. isHexidecimalDigit():

This will reveal whether we work with hexadecimal or not.

8. isLowerCase():

This one will reveal whether the character is lowercase or not.

9. isPrintable():

This will determine whether the character is one can be printed off using the console.

10. isPunct():

This will reveal whether the character we get back is an punctuation mark.

11. isSpace():

This one will reveal whether the person we're returning is a spaceman or not.

12. isUpperCase():

This one will determine if the character we're receiving is an upper-case letter.

13. isWhiteSpace():

This will determine if the returned character we're working on will be whitespace, which is similar to an elongated line, space or a tab.

Handling Our Random Numbers

It is also possible to work with certain random numbers.

Random numbers likely to be the functions we can employ to create random numbers for every program we create.

It is important to note in this regard that computers can't actually be completely random or spontaneous, even though.

This is due to the fact that all of the components we can view are dependent on the input that the computer receives and will be able react in a manner that is most rational to it.

The initial type of random number we will be able to utilize in this case will be used to perform the function of randomSeed(number).

This will start by generating the random number to allow us to use that number to work.

It is first necessary to ensure that we feed in a number throughout first, then it will begin again at random where the computer program chooses out of the order in pseudo-random number generator's sequence of numbers.

It is possible to go a step more in this direction and use the role of random(OPTIONAL minimum and maximum).

This one will be beneficial because it can be connect with the generation of a random number.

The highest value is likely end up being the highest number you're allowing here, and the lowest number is the smallest amount you are allowed to allow.

If you don't follow through and define the minimum you wish to use, we assume it is the level of 0.

A look at Bitwise Functions

The final aspect we'd like to examine in relation to the capabilities that are part of Arduino API Arduino API are the bitswitches functions.

They are a particular kind of function designed help you use bytes as well as the parts of your code.

They will be regarded as the smallest bits of data that computers can actually work with.

There are some arguments that there are other smaller data types, but for practical needs it is the case that we should stick with the most compact parts.

Some of the many bitswise functions that we're capable of using when using Arduino include:

1. bit(bitNum):

This one will reveal the value of a certain bit.

2. bitClear(variable, bit):

This will be the method we use to assign the specified bit of a numerical variable to zero.

3. bitRead(variable, bit):

This will provide the part of a specific numeric variable.

4. bitSet(variable, bit):

This will set a variable's bit in the same manner as the position indicated by bits to 1.

5. bitWrite9(variable bit, bit, 0, or 1):

This code will change the variable's bit at the specified location within the variable. It will be set at 1 or 0 in accordance with the value you specify inside the script.

6. highByte(value):

This will give us the top amount of bytes of a particular value we can find.

7. lowByte(value):

This will be opposite to the one before and is expected return the lowest bit of the number you select to analyze.

As we have seen in this article, there are many different features that we can to utilize in conjunction with the API that comes with Arduino.

Knowing what each of them can do for us , and studying them more deeply will greatly impact the things we can accomplish with them as well.

When you become more acquainted to your experience with the Arduino control and the multitude the cool things it can accomplish and accomplish, you'll quickly understand the reasons why it is an excellent choice to use and be able to observe these functions performing actual tasks too.

Chapter 11: Stream Class And Arduino

It is now time to look at an entirely new subject that we can look at in more depth as well.

The next chapter we're going to be working with the stream class which is different from the work we've done in the past.

However, if you're looking to use strings in our programming and programming, we must be aware of exactly how stream classes is likely to function for us.

It is evident it to be a simple concept to use and is particularly so once we get through it more.

Although it is simple, however we must ensure that we understand how crucial it is to be working with.

The stream class which we are working on it in its own will be based on the information we have gathered from our chosen source, and following this process as the way to create your sketch to use.

Because the stream is going to be primarily about reading the data it is important that we discuss the use of the mouse and keyboard in this type of course, though they might not directly relate to the subject.

Connect the keyboard and the mouse right now, as they'll make the entire procedure a bit more simple.

If you choose to use this information, particularly when you have all the data prepared and ready for use and you are ready to go, you will notice that there will be times that the sets of characters we'd like to work with will be quite long and include enough characters to form an entire sentence.

The idea behind the string from the strings is to assist us in doing this task on the board.

Strings are likely to be a set of characters which are connected in a way that is comparable to an array.

That means that they will be considered to be contiguous within the computer's memory and when computers take an examination of how it works, they will view them as one large and interconnected unit.

The strings are a part of the instrument. Working with them will simply mean that we will need to spend the time to study how to manipulate these strings, and also the capabilities of the board that will enable us to accomplish this.

If we only take a look at the simple nature that the strings have, it's simple enough to comprehend.

Strings will only be what we call character arrays.

This implies that we are likely to continue to see these strings to appear as if they're component of C language, and should you have knowledge of the C language, then you'll be able to see how easily it is to use.

Be aware that the strings in this case as a result of this are likely to require a lower-level abstraction that we have to deal with.

For instance, in many of the modern programming languages the strings aren't actually going to be visible in the character , but rather as an array.

Instead of using them instead, they are likely to be considered an abstract thing instead.

Even if they're seen as characters in this way, they'll still be treated as such.

Chapter 12: Designing Our Own User-Defined

Functionalities

The last thing we must examine when we are coding within Arduino's Arduino board will be how we can use the user-defined functions.

In every language that is coded there will be functions that are included.

We can take them out at any moment that we'd like make use of them. They are ours to use as we'd like.

However, there are instances when we need to write your own function, custom functions that can manage a lot of the tasks that we require but are specific to the particular code we're working to create to run our program.

The second type of functions are likely to be called user-defined functions.

Functions will be crucial in the code that we intend to code in our language, and in other languages, too.

They're good at making sure that the code is clean and organized and we're in a position to reuse parts of the code, if we would like.

It's even designed to assist us in ensure that the code will be able behave in the way we'd like it to all the way.

In essence the functions will turn out to be merely tools we can to utilize, and which are designed to support the specific actions that we wish to accomplish within our codes, much like we might guess in the first place.

While we've actually gone through the codes within this guidebook, which is expected to contain functions inside these codes, it's time to explore the specifics that are included with thesecodes, so that we have an comprehension of the way they function and why we'd like to utilize these codes as well.

This will help us to discuss some of the features that we might not have considered and did not understand when we first started because we didn't talk about the functions in beginning.

Let's take a look at the various functions and see what we can do using this.

The first thing to do involves taking a close examine how we can define any of the tasks we have on our list of things to be working with.

This is a crucial process when working using our own functions that we have defined by the user We must be sure that we're following the correct way on our software.

An excellent example of the codes that we could use to accomplish this would be:

Employee earnings based on floats (float hours worked, float payrate) {•

Results in floats the following: // this is the value we are returning when the function has been called up. We must ensure that the result will be the same kind of data we will use before the function's name.

Results = hours Worked * payrate

Return results/return informs the program that it has to return a value at the point it was first named.

}

While you look at this then, open the compiler you're using on Arduino and start entering it.

Also, you should look at the code and consider what it actually has to provide in terms of what we can accomplish using it, and the things you already know from the code, based on various other subjects that we've discussed in this book.

It might surprise you how many other elements you'll come to recognize based on the information we've been able to discuss in this book, and the things we've done in the past.

The above function will be an excellent one to use since it can handle two arguments and needs to be able to handle this to finish its work effectively.

There will be two sides, including the hours worked and the pay rate the company must manage in order to ensure that things are handled.

It can also be able to take some time to work on a little bit of the basic math we discussed in the past, prior to bringing up a floating number to determine the value from the numbers present.

The amount we'll receive from this program will assist us in completing or stop our program, and then return the value that was given to us and this should occur right after using the word return which will become a variable based on the results of the calculations we made prior to.

This will have to be done in order to assist us comprehend how these functions are intended to function and how we can perform some work using these functions.

It may appear like too much details, but when we have all this in mind, it's the right time to look through and dial on all the functions.

This will help us to understand how much the earnings of employers are. It will often help us understand the process for us to understand what happens in this procedure as we go through the process and even incorporate certain of the functions defined by users.

To make some sense this and understand what we're trying accomplish, we should apply the following code:

```
void loop () {

floathours worked = 37.5;

float payRate = 18.50;
```

The result of a float is employee earnings (hoursWorked and payRate)

The final Results will be 693.75

It's a simple code that is easy to use, however it is sure to demonstrate what we are capable of doing in this case, and the reason why each of the components are crucial along the process.

It's also an easy way to gather ideas about how the function will work in the long run.

When we arrive here, the very first thing we'll see in all of this is the function we must declare and ensure that it is declared to be independent of and independent of all other functions.

This means that we have to have the time to write the code for every function that we wish to develop and that it must be executed either within the setup function or in the loop function we are working with.

Chapter 13: Operators To Use With The Arduino

Board

In the case of working with C as a C language, it's essential to at least some time studying the operators available in the language.

There are operators in all the programming languages you could want to devote your time with Although they may seem like a basic part of code, they're going be crucial to some of the work you're trying to finish.

They're small, but they're powerful and, without them, you won't be able to make your code perform.

These operators, both in the C language, as well as in various different languages are symbolic symbols be used to aid us perform operations that aid in accomplishing things as quickly as it is possible.

The operators we're going to utilize when we are focusing on this language could include options such as bitwise, logical and arithmetic in addition to Boolean.

Although all of them are essential however, there are additional operators can be focused on, we're going focus on the two operators you are most likely to come across when you work with the Arduino controller: the arithmetic and Boolean operators.

Let's get started and find out how this will play out.

The Arithmetic Operators

The very first type of operator we'll be capable of working on will be algebraic operators.

These will be the operators we should focus on to be able to manage the mathematical calculations we're dealing with.

If you're writing programming and you have to add or subtract the sum of two figures (or greater) from one another, these operators for arithmetic are likely assist you.

The arithmetic operators will be able include subtraction, addition multiplication, as well as dividing sections of the code according to what you'd like to see happen with different numbers that are in the code.

It is also possible to have multiple of the operators for arithmetic show in a single section of the code you're working with or include multiple operators on the same line.

It is merely a matter of remembering that we need to follow the concept of operational order.

This will ensure that we are capable of going through and complete the math correctly and we'll be able to come up with the correct answers in the process.

The Boolean Operators

Another alternative that we are in a position to work with within this language will include some Boolean operators.

They are founded on the notion that the result you're getting will be accurate or not.

There will be some different options we can work using when working with Arduino. Arduino system, for example and && as well as !.

These are the two primary kinds of operators that you should concentrate on when working with this type of board, and getting it to work for certain results you want to achieve during the process.

Understanding how to achieve these and making sure you can see them appearing within the code correctly and at the correct moment, is crucial to the overall programming we're doing.

Chapter 14: Computer Interfacing

Now is the time to look through and perform some of the tasks in connection to Arduino. Arduino system.

There will certain situations in which you'll want to grab this controller and hook it to your computer to complete your work you'd like to accomplish.

The interface between Arduino and computers is expected to be widespread yet there are going to be some steps we're able take to make sure everything goes smoothly.

The concept of connecting your computer with your Arduino is likely to be the ideal option to work with when you wish to connect your Arduino to your computer.

However, it is also going to be contingent on what cables you'd like to use, or what's available in the moment.

While we're here, it is important to keep in mind that each controller of Arduino which we use is capable of an easy connection to the computer, if we choose to take the time to connect to an USB port.

Connecting the Arduino to the computer you choose will usually be dependent on the language of programming we intend to use as well as depend on the additional components that must be incorporated on this controller.

All of this and more can affect how well the Arduino controller will be able to connect to your computer.

As we can see it is likely to be a process that's somewhat difficult to understand and is not always simple for someone who is new to work through and understand.

This is why we'll spend the moments in this section to study computer interfacing and what it entails to ensure that we can have a greater chances of making it work using our Arduino controllers. Arduino controllers, too.

The FTDI Chips

One of the aspects we must consider when we talk about computing interfacing, is the concept of FTDI chips.

The Arduino boards you could use will be able of sending as well as receive information, making the data transfer to the computer with the aid of an USB port however there are two major variations.

The exceptions will come from the Mini boards as well as the Lilypad alternatives.

However, even with these exceptions, you'll be capable of connecting these boards to your computer and have it communicate the way you desire using an FTDI interface.

This will basically be a small device that can be used to in the exchange of information between the computer as well as Arduino controller. Arduino controller , in whatever way that you'd like.

When you are doing other activities that you could do you can try to use the Arduino controller to examine and read certain values from the sensors, such as the light and temperature which you then observe these values by the LEDs present on the board.

In this chapter, we will need to spend some time to look at another optionthat will be referred to as serial interfacing. This allows us to extract measurements from the sensors, and then transfer them to the computer in order to help us figure all the data that we require.

This is why we will look at an example of how we're capable of working with temperature sensors using the aid using the serial interface.

We must ensure that we have some items in the back of our minds prior to time, like the Arduino UNO boards the breadboard, and an temperature sensor as well as the USB cable.

The code we are going to need to create to make everything take place.

The codes we will employ here will consist of:

Const Int SensorPin const = A0

int reading

```
float voltage;

Temperature of the float;

Setup not in place()

( Serial.begin(9600);}

Void Loop ()

}

Reading = analogRead(sensorPin);

Voltage is Reading * 5.0/1024;

Serial.print )voltage);

Serial.println("volts");

temperatureC is (voltage minus 0.5) * 100

Serial.println("Temperature is");

Serial.print(temperatureC);

Serian.println("degrees C")'

Delay(1000);

}
```

Once we've been able to confirm and then upload the code we wish to work and click on the Serial Monitor that comes up.

You will notice that there is a menu going to display the temperature sensor readings.

We now want to utilize any source of heat we can locate to increase temperatures of our device.

Remember that with this equipment, we're capable of handling temperatures as high as 150 Celsius.

Also, you may notice the (--) symbol appear.

This doesn't mean we're working with negative numbers in this case It's just an error in programming that is temporary.

While we go through this, we're capable of looking at other components to understand how we were in a position to make this be able to meet our requirements and help us write our code.

For the first step first, we must examine our Serial.begin(9600) section.

This will be the document we will write down to ensure that we have set up the required communications with Arduino controller Arduino controller and our computer, making use of the USB connection and the cable.

This allows the two parties to share and receive information to one other.

Furthermore it is also possible to deal with two key variables that must be included in our code. this includes the TemperatureC and voltage. Both of them must be defined using a floating point instead of an integer.

The reason is that the device which is able to change to measure temperature will need to be precise and the outcome must be a floating value so that we can achieve the full precision, not an integer that is able to work with whole numbers and requires us to round it.

From this point, we need to move on to the next part of our code, which is going to be the part for reading = analogRead(sensorPin);.

This is the procedure we will need to follow to capture the input analog onto the correct pin of the procedure.

Conclusion

Arduino is an open source platform for electronic prototyping that is based on a flexible and simple to use software and hardware. It is geared towards creatives, designers as well as amateurs and those who is interested in designing interactive environments or objects. Arduino can sense environmental conditions via signals from sensors, and then interfacing with the surroundings, controlling motors, lights, or other devices. The microcontroller on the PCB can be programmed using the Arduino cable-based programming language as well as the Arduino environment that is based on processing environment. The projects created using Arduino may be autonomous, or can communicate with a computer to accomplish a task with specific programs (e.g., Flash, Processing MaxMSP).

The technology developed in 2005 by Italian Massimo Banzi and other collaborators, was created to facilitate the classroom teaching of electronics to learners of design as well as artist.

The platform was designed to be affordable since students had to build their ideas while spending the least amount of money. It's easy to understand not it?

Another fascinating aspect of the project was the idea to develop an open-source platform accessible to the public which was a huge help in the process of its creation.

What are the possibilities?

In the event that we link input devices motors, electronics, sensors and displays, antennas or other devices to the Arduino the possibilities could be limitless.

The applications it has are numerous, from amusement to home automation, art and even helping others. For instance Arduino Arduino was previously employed to build an electronic beer cooler that was that was controlled by an iPad. the user could track the flow of drinks and gather information about various types of beer. The plate was also used to tell the temperature as well as reveal who had drank the most. Another team developed an e-sensible glove that allows visually impaired people "see" obstacles on the path. Similar to this an additional user developed an outfit that uses an Arduino LilyPad variant (designed to make design wearable) that is part of Arduino that tells you that cyclists are changing paths by using LEDs on the inside part of the jacket.

This book discussed some amazing projects using Arduino to demonstrate that this is a device that assists all those with a creative mind.

You can create LED cubes to create three-dimensional images, hidden knock detectors. After all, security never gets enough, and sneakers that tie your shoelaces for you, such as "Back to the Future 2" In fact, the initial prototype was worked enough that it ended up becoming an actual product, and was made available to the commercialization ...

Accessible Technology >> Movement Maker

With already got an Arduino board is a good start, but the next procedure to get started using it is to download the development environment from the PC and link it to the board using an USB cable.

Affordable technology, such as Arduino and 3D printing is essential, not just for testing or 'homemade creations.'

Most people know that the makers of Pebble - a device that made 10 million dollars during its crowdfunding effort - could not secure a second round of financing, and they went on the public (Kickstarter) to raise money. But what is less known is that they utilized Arduino to build their prototype.